Air Conditioning:
Impact on the Built Environment

also of interest

Heat Pumps for Buildings
edited by A.F.C. Sherratt

Energy Management for Buildings
edited by A.F.C. Sherratt

Steam Trapping and Air Venting
L.G. Northcroft and W.M. Barber

Applications of Heat Pumps to Buildings
edited by A.F.C. Sherratt

Air Conditioning: Impact on the Built Environment

Edited by A.F.C. Sherratt

Hutchinson

London Melbourne Sydney Auckland Johannesburg

Hutchinson Education

An imprint of Century Hutchinson Ltd
62–65 Chandos Place, London WC2N 4NW

Century Hutchinson Australia Pty Ltd
PO Box 496, 16–22 Church Street, Hawthorn,
Victoria 3122, Australia

Century Hutchinson New Zealand Ltd
PO Box 40–086, Glenfield, Auckland 10, New Zealand

Century Hutchinson South Africa (Pty) Ltd
PO Box 337, Bergvlei, 2012 South Africa

First published

Library of Congress Cataloging-in-Publication Data

Air conditioning.
"Contains the papers presented at the conference of the same title organised in March 1986 by the Construction Industry Conference Centre Limited in conjunction with the Chartered Institution of Building Services Engineers, the Royal Institution of Chartered Surveyors, the Institute of Energy, the Institute of Refrigeration, the Department of the Environment, and the Department of Energy of the United Kingdom" — P.
Includes bibliographies.
1. Office buildings — Air conditioning — Congresses.
I. Sherratt, A.F.C. II. Construction Industry Conference Centre.
TH7688.04A37 1987 697.9′3523 87–7686

ISBN 0–89397–277–0

Printed in Great Britain by St Edmundsbury Press Ltd
Bury St Edmunds, Suffolk

British Library Cataloguing in Publication Data

Air conditioning: impact on the built environment.
1. Air conditioning
I. Sherratt, A.F.C. 697.9′3 TH7687

ISBN 0–09–173029–5

First published in the UNITED STATES OF AMERICA in 1987 by Nichols Publishing Company, P O Box 96 New York NY 10024

Contents

Acknowledgements

This book contains the papers presented at the Conference of the same title organised in March 1986 by the Construction Industry Conference Centre Limited in conjunction with the Chartered Institution of Building Services Engineers, the Royal Institution of Chartered Surveyors, the Institute of Energy, the Institute of Refrigeration, the Department of the Environment and the Department of Energy.

Organising Committee

Chairman: A.F.C. Sherratt, BSc, PhD, CEng, FIMechE, FCIBSE, FInstR

P.S. Howes, FCIBSE
(representing the Chartered Institution of Building Services Engineers)

R.W. James, BSc, MEng, PhD, CEng, MIMechE, MCIBSE, MInstR
(representing the Institute of Refrigeration)

D.J. Martin, BSc, PhD, MInstP
(representing the Department of Energy)

C.J. Swan, BSc, CEng, MIMechE, MCIBSE
(representing the Department of the Environment)

Deryck L. Thornley, FCIBSE, FHKIE, MASHRAE

R.B. Watson, FRICS, MACostE
(representing the Royal Institution of Chartered Surveyors)

G.G. Webb, BSc, MInstE, CEng
(representing the Institute of Energy)

1 Air Conditioning — An Ever Expanding Market

D.S. Gillingham

Introduction

Worldwide, air conditioning is unquestionably an expanding market, principally because of continued growth in developing areas such as South East Asia and to meet specialised industrial needs. In particular developed countries, it is less easy to comment as commercial and industrial activity is much influenced by local economics and financial policies, with consequent effect on air conditioning demand.

The debate is increasingly on how well air conditioning should be done, in relation to the balance between capital and running costs, rather than whether air conditioning is needed. In what follows, the more important sectors of the market are considered with a review of the trends and developments occurring, largely based on recent experience and the views of knowledgeable people.

Office Blocks

The building regulations and thermal performance

Part L of the 1985 Regulations (1) has the principal aim of reducing heat loss by prescribing lower overall U-values and is to be much welcomed. The consequences on the building and its systems may not be immediately obvious but are important, nevertheless.

1 The sensible heat gains through the envelope of a typical, west-facing module with 35% single glazing (as an example) are now only about

70% of those through a similar envelope with 50% single glazing, before Part FF of the 1976 Regulations (2) was applied. Accounting for typical internal gains reduces the supply airflow from 100% (pre 1976) to about 80% now. Obviously the sizes of plant and air distribution ducts are reduced.

2 The outside temperature at which heat loss balances heat gain is lower. For the simple example above it falls from 7.0°C to 1.7°C. Buildings therefore need to be cooled for a longer period of the year and a system maximising the natural cooling potential of the outside air (the refrigeration plant being off) will be more economical to operate than one that does not.

3 Lower fabric heat losses reduce building cooling at night in summer with an increase in the energy for refrigeration next day. To minimise this the case for accelerated night-time cooling, by running the system fans, is worth considering.

4 Infiltration is a larger part of the heating requirement with the improved U-values and system performance will be more sensitive to changes in wind penetration.

5 Slower cooling down and heating up times will reduce the energy savings possible by intermittent heating.

The use of computers and other office equipment

Computers

The trend is to have central data bases with communication networks to personal computers, not necessarily in the same building. Laser discs are superceding magnetic discs for storage and fibre optics probably replacing electrical transmission to give quicker responses and save building space. The trend is to use unmanned, central computing equipment in unlit rooms with typical conditions of 20°C to 26°C, 40% to 60% humidity and virtually no restraints on noise or air movement. Major heat-producing equipment is directly cooled by liquids, reducing the space and energy needs for air cooling. In line with this, computer capacity is increasing without a growth in computer size.

Humidification is still a likely need to counter electrostatic problems but possible developments in materials and electric bonding techniques may alter this.

System design must reflect these changes, with minimal fresh air needs, a very clean environment and no significant increase in capital and running costs or in plant space.

Office equipment

Typical equipment in use covers a wide range: VDUs, printers, typewriters, telex, fax, table lamps, calculators, graphic plotters, copiers and private computers, often supported by an uninterrupted power supply. Currently, there is a transition from the use of paper to electronics as a form of communication and recording. These methods are in tandem at present but presumably it is only a matter of time before paper, as a means of communication, largely disappears (3). Typewriters, printers and copiers will greatly reduce, the number of VDUs will increase and task lighting will augment a reduced background illumination level.

A traditional allowance of 5 W/m^2 to cover heat gains from office machines has increased to 20 W/m^2 in the last ten years (4). This level might be a plateau but special areas, such as dealers' rooms, may have much higher values and need supplementary cooling and electrical services.

In spite of this, it is interesting to note that many offices, designed to cope with 5 to 10 W/m^2 are surviving in the dawn of the office electronic age, without major changes to the air conditioning system. The reasons for this may be:

- Air conditioning systems have surplus capacity.
- The load factors for office machines may be small.
- Temporary increases in local temperature may be tolerable.

Services designers must cater for future, local, increases in heat gains by providing adequate system flexibility and building designers must let them have the increased space to permit this (5).

The office of the future

The relationship of people to buildings and the new information technology is described in the Orbit (Office Research on Buildings and Information Technology) Report (4). The office of the future will be designed to contain equipment ranging from desk-top micro-computers to larger machines, such as digital PABXs with specialist facilities for data processing and satellite communication. It is predicted that the impact of information technology will increase the decentralisation of office organisation, most buildings having multiple tenant occupation in smaller units of accommodation (100^2 to 1000 m^2). The trend may be to deeper buildings, with cellular peripheral offices and cores containing equipment. The Orbit Report believes the popular concept that equipment for information technology can operate in conventional office areas is optimistic, because of the requirement for machine rooms to house computer equipment. Changing loads and needs will also demand system and building flexibility in use; although traditional comfort cooling systems will cope with general areas the report believes that gains at work-stations could reach 40 W/m^2. To provide this flexibility, the additional capacity built into the sizing of terminal units will considerably increase the space required to distribute the services throughout the building.

The architectural press generally feels that many of the speculative office buildings erected in the 60s and early 70s have services which are inadequate for the new technology and that the cost of adaptation would exceed that of new construction. It is doubted whether renovations can ever be justified on cost grounds but, without renovation, some buildings will become dinosaurs of greatly diminished value.

An internal study (6) of office accommodation used by ten oil companies in Aberdeen gave a practical view of the impact of information technology. It was concluded that the heat gains resulting from the use of electronic equipment could only be dissipated by mechanical cooling, both in open plan layouts and in peripheral offices with electronic machines. It is evident that, even in a cool northern climate, natural or mechanical ventilation is inadequate where information technology is in vogue and air conditioning then becomes necessary.

The office of the future will require a highly flexible air conditioning system, probably variable air volume, using plants installed on a floor-by-floor basis. Cooling towers and plate-heat exchangers on the roof will feed clean cooling water downwards through the building and branches at each floor will distribute the cooling water to the local air conditioning plants. Supported floors will become fairly universal, to accommodate the electrical and mechanical services. Individual tenants will be able to operate their own air conditioning systems to suit their working hours, irrespective of the rest of the building and, in this way, the air conditioning systems in a multitenanted office block will be operated more effectively and economically.

Decentralisation

For some years, under the pressure of rising rents and rates, there has been an apparent tendancy for tenants to seek office accommodation away from city centres. Press comment suggests that decentralisation is increasing but it appears (7) the annual rate of decentralisation associated with large moves from London between 1979 and 1982 is less than half the rate in the preceding fifteen years! This does not mean it has ceased: from 1966 to 1976 office employment fell at a rate of 1% per annum (8) and, since then, it has continued to decrease in the City (9).

The main reason for moving away from the centre has been operating economy. Rents are high in central London, so are rates and service charges which have played a part in encouraging decentralisation (10). Total annual costs of accommodation (rent and rates only) are about £550/m^2 in the City, £130/m^2 to £190/m^2 in outer London, and £50/m^2 to £200/m^2 in the provinces. The tendancy seems to be for re-location fairly close to central London, either in the green belt or the suburbs (11).

Due to electronic machine rooms, air conditioning continues to be necessary even in decentralised offices. The increased use of VDUs requires good lighting without glare and good solar shading. It is then less possible to open windows for natural ventilation in warm weather.

Tenants want new, well-appointed buildings and those with only basic services are difficult to let, particularly so in the southern and south eastern areas outside London (11).

The Retail Sector

It has been argued (12) that retailing is possibly the most competitive business today and that success depends on an acute awareness of the changes occurring in our society and a rapid response to them. Consumers' attitudes are in a state of flux and High Street shopping reflects such changes, one of which is an increased expectation of quality. Over the last ten years retail expenditure has increased steadily to a current annual value of £27 billion in the grocery trade (13). The increased competitiveness among traders has changed retail methods with a rise in the total trading floorspace, principally by an increased number of larger stores. Although new construction work for shops and stores was steady at an annual figure of about £540 million from 1980 to 1984, superstores and hypermarkets have had an increasing share (14). This is because they satisfy the shoppers' growing demands for adjoining car parks, more competitive pricing, and a visually pleasing and comfortable environment inside the store. Sainsbury claim that the average customer purchase in their newer, larger stores is three times that in their older supermarkets (15).

Superstores are defined by the Institute of Grocery Distribution as having at least 2,300 m² of retail area and hypermarkets as having twice this. In both 1983 and 1984 the number of superstores in the UK increased by twenty-seven (16) corresponding to an annual increase of 8.5% (see Table 1.1).

Table 1.1 *Superstore numbers by operators*

	1982	1983	1984	%change 1982-84
Argyll Foods	11	12	16	45.5
Asda Stores	65	71	77	18.5
Dee Corporation*	12	12	13	8.3
Fine Fare	31	35	39	25.8
Hillards	7	8	7	--
International (now part of Dee)*	18	18	5	(72.2)
Wm. Low and Co.	2	2	2	--
Morrisons	14	17	22	57.0
Safeway	1	1	1	--
J. Sainsbury	10	10	19	90.0
Sava Centre	5	5	5	--
Tesco	70	79	83	18.6
Woolco	11	11	11	--
Co-operative societies	47	51	63	34.0
Independents	14	13	9	(35.7)
TOTAL	318	345	372	17.0

Company reports indicate that this trend will grow over the next few years and this is supported by the fact that there are about 500 shopping centres in the UK with at least a further 45 under construction (16).

The probability is that trading for 7 days/week will become common in both superstores and shopping centres and the move to better quality environments for customers will continue as the large grocery companies compete for an increased market share. As an example, Toys R Us, the biggest specialised toy retailers, are building large, windowless stores with air conditioning. The major retailers recognise that the provision of good

Table 1.2 *Stores growth*

STORE	YEAR	NEW STORES	NEW AREA SQ.FT.	CAPITAL EXPENDITURE £ MILLIONS
Tesco	84	11		
	85	15	500,000	
	86		"	
	88		"	
Safeway	84	7	200,000	
	85	7	200,000	40
Asda	84	8		62
	85	12		
Gateway	84	31	381,000	
	85	31	381,000	
Carrefour	86	3		
Marks & Spencer	84		245,000	
	85		330,000	240
	86		330,000	240
Sainsburys	84	15	391,220	246
	85	15	407,810	

environmental conditions is essential if the customer is to spend any length of time in a retail store: heat gains from high lighting levels, electrical items of equipment and people, make air conditioning mandatory. No superstore owner would contemplate planning a new store or refurbishing an old one without installing up-to-date air conditioning.

Systems of air conditioning currently being installed must reflect the latest technology, to suit the building and its environment. In particular, the running cost must be carefully considered because it is a relatively large part of the owning and operating cost of the building and its service. As a result, air conditioning systems are incorporating heat pumps and recovering heat from the air discharged at bakery ovens and the like.

A typical annual distribution of the energy used in a store selling general merchandise and food is given in Table 1.3 (17).

It is seen that the energy for cooling and moving air amounts to 27.9% of the annual energy bill. This is the reason why Tesco Stores Ltd., in collaboration with the Energy Efficiency Office, have set up a two-year exercise to monitor and target their consumption of energy. They estimate that 10% saved in energy costs will equate to an increase of 4.5% in net profits.

Table 1.3 *Energy used in stores*

Item	% Energy used annually
Air cooling	12.5
Air movement	15.4
Heating	14.1
Hot water	6.0
Sales lights	21.9
Non-sales lights	6.7
Food chilling	7.4
Catering	8.0
Lifts and escalators	4.5
Miscellaneous	3.5
	100.0

If this were applied to all retailers, the annual saving in energy would represent £70 millions.

Hospital buildings and health care

The development of air conditioning generally in hospital buildings originates from work done on the importance of controlling the spread of air-borne infection (18). Whilst this is still a primary requirement, it is not in isolation a prescription for air conditioning; indeed, in the case of a poorly designed system, the reverse could be true (19).

Air conditioning, as distinct from good ventilation, can be essential for clinical reasons to maintain specific conditions for the well being of particular patients, or to provide comfort that is otherwise unobtainable, or for both purposes (20). These principles are well understood and form the basis of current good practice in the design of hospital services. Typical examples of air conditioned spaces in UK hospitals are: operating theatre suites, intensive care facilities, delivery and special baby care units, sterilising rooms, pharmacies and pathology departments.

Dual-duct systems, quite widely used in the past, have given way to single-duct schemes in the interests of lower costs and greater efficiency in the use of energy. The trend is for a looser fit and a flexibility to respond to change.

The extent of air conditioning for strictly clinical reasons is comparatively small, most installations being for comfort, and there is little evidence of any immediate change. Within the whole building, the air-conditioned fraction is not large but the proportion may be higher in private hospitals because of the concentrated surgical facility which is often a feature of them.

There has been a change in the planning of hospital buildings, leading to the 'Nucleus' designs on which most work in the public sector is now based (21). A consequence is that future air conditioning demands in hospitals are more predictable, an up-turn in the UK appearing unlikely. Furthermore, the increased concern with energy consumption in hospitals may also limit future expansion. Energy studies commissioned by the DHSS (22) relate to NHS hospitals but private hospitals are no less motivated in reducing controllable energy costs. Given cost restraints in both sectors, the emphasis in the UK is not on expanding the scope of air conditioning but rather on re-defining and minimising the need by better planning and design in buildings.

Any expansion of air conditioning in the UK is not likely to come, at least in the short term, from increased scope but rather from the construction of new facilities. To some extent this has already happened in the private sector but whether this will be sustained and what the future holds is not immediately obvious. On the other hand, there is a major growth market in the developing world — there have certainly never been as many hospitals under construction worldwide, in countries where air conditioning is even more vital than in temperate climates.

The inference is that the expansion of air conditioning is not to be viewed purely in terms of size. Equally relevant is its growing importance.

Laboratories and animal rooms

By focussing attention on the possible dangers to which workers in laboratories may be exposed, the Health and Safety at Work Act has had a considerable impact on the design and operation of research laboratories and the services within them.

One aspect of this is seen in the routine use of exotic, synthetic compounds. Their totally unknown health hazards, together with the more general use of radio-active compounds, have taken experimentation from the laboratory bench into fume cupboards of ever increasing size that require vast amounts of make-up air. A new draft British Standard (23) proposes minimum velocities for fume cupboards with permissible deviations that will present many problems to the design engineer: very high room air change rates have to be accommodated and precise attention must be paid to the airflow pattern in the vicinity of the sash openings on the cupboards.

With biological research likely to involve the use of small animals for some time to come (until alternatives have been developed), the problems of cross-contamination between animals and humans are well recognised and remain. Allergies (such as asthma) are medically notifiable conditions and screening laboratory workers for them is now commonplace. Air conditioning makes a valuable contribution to human health in this respect by giving carefully controlled airflow patterns in animal rooms. It is to be noted that, in this field of research, consistent results can only be achieved when the laboratory animal is not subjected to stress, often caused by environmental factors such as temperature fluctuations, excessive air movement, noise and so on. A recent Government White Paper (24) proposed that the Home Secretary should have powers to issue guidelines and codes of practice on scientific procedures. These will provide detailed information on animal husbandry and it is evident that air conditioning must play an important role in satisfying the requirements.

Research laboratories are not immune from the lawless trend in our society. The cost of basic research is now so great that there is an increasing temptation to industrial espionage, imposing and need for greater security precautions in laboratories. Furthermore, where animals are used, there is the requirement to protect companies from the more violent members of activist organisations. These matters are turning laboratories into highly secure buildings with monitored entry systems. In its train this brings the need for non-openable windows, making air conditioning essential.

New industries

Industry in general

As the base of manufacturing in the UK changes and there is a movement towards production requiring a higher level of technology, the need is apparent to apply an improved standard of environmental control through all stages of the manufacturing process.

In research and development, often sadly underfunded in the UK, the following trends are clear:

1 Team research, involving larger and more open space, is becoming the norm.
2 The necessity for mechanical and electrical services within the space is increasing substantially.
3 The use of computers and micro-processors has become an everyday part of the activity and will increase.
4 Rising energy costs are identified as an increasing fraction, emphasising the need to find new solutions that are efficient in their use of energy.
5 The quality of the environment is important if high calibre staff with the necessary talent is to be attracted.
6 Shorter product life-cycles and quick obsolescence require a large degree of flexibility because of the rapidly changing use of the space.

In the production stage many of these influences also apply. Additional factors are: more automation with expensive machine tools, computer and micro-processor controls, cleaner environments etc. As more attention is paid to health and safety at work, the need to exercise more refined control over airflow paths, relative air pressures, hazardous materials and fume emission will make evident the increasing importance of air conditioning.

For warehousing and product distribution, the introduction of advanced technology in the form of the computer-controlled, automated warehouses is having a significant effect. High value, constantly changing stock requires proper attention to environmental control. Nowadays, even in such works offices an upgrading of the everyday working conditions is likely to be necessary to attract staff of the appropriate quality. Fatigue of the kind resulting from studying VDUs imposes new environmental standards.

The new clean industries

The first generation of clean industries has established itself in parallel with the decline of the older, dirty, so-called 'smokestack' industries. Clean rooms for electronics, sterile suites for pharmaceutical manufacture and hospital operating theatres all use air conditioning. As the new industries search for novel, more effective processes to retain their leading position, so the more conventional industries seek to improve their own performance. By working in and being stimulated by a wide range of industrial situations of this sort, the air conditioning specialist can acquire and apply techniques for the benefit of them all.

The imposition of regulations and directives introduce other aspects of industry where the manufacturing environment is of vital concern. For example, any pharmaceutical manufacturer wishing to sell his products in the United States must satisfy their Food and Drug Administration that stringent validation procedures (frequently involving HVAC systems) can be met. Increasingly, also, EEC regulations for food production and animal research will require a more critical control of the environment.

The impact of the micro-electronics revolution has been felt by the air conditioning engineer: the demands of research workers (and manufacturers) in this field are for standards of cleanliness with stable control over temperature and humidity that were unheard of a generation ago. With research involving dimensions in the sub-micron range, the art of the air conditioning engineer is taxed to the full. As micro-electronics moves into the next generation of devices, with circuit components measured in molecules rather than microns, the ability to control temperature, humidity and cleanliness, within limits not yet attainable, will be essential. To compound this, the use of highly toxic, spontaneously inflammable and explosive gases demands the most rigorous attention to exhaust ventilation systems and make-up air.

Elsewhere, our increasing reliance on nuclear power as conventional fossil fuels deplete introduces the storage and containment problems of radio-active waste, requiring us to be absolutely sure that all HVAC systems fail safe as they increase in size and complexity.

Airport buildings

Air travel is now firmly established as a conventional mode of mass transport and requires a brief mention, although the subject of services for airports was recently dealt with thoroughly (25). The importance of air conditioning to create a comfortable environment for the vast numbers of passengers handled merits comment, for were it not for the benefit conferred by air conditioning, modern airport terminals would be virtually uninhabitable in temperate climates and literally so in warmer parts of the world.

Building management systems

In accepting the challenge presented by the use of heat pumps, variable air and water flow, better control response with greater accuracy and the introduction of energy management systems, the controls industry has adopted new technologies from computers and communications to satisfy the needs of HVAC systems. In employing such new techniques the air conditioning designer should be wary: the economies and refinements offered must be weighed against the possible difficulties of installation or operation, and even the loss of convenient manual adjustments that used to be available.

Although it is impossible to survey accepted recent changes fully here, the greatly increased control flexibility following the introduction of the micro-computer as the basis of monitoring system design, can be noted (26). The evidence is to be seen in any recent, major, services installation where wall-to-wall panels have been replaced by desks with colour VDUs and operator keyboards.

Less visible are the technological changes making this possible and offering energy management facilities hitherto impracticable. The transition has been from passive monitoring, probably with separate controllers in the one panel, to taking advantage of the opportunities provided by the use of centralised micro-computer control, or the use of central control plus so-called, intelligent out-stations. It is also possible to use bureau services to supervise and co-ordinate the operation of multiple installations connected by telephone modems (27).

A central computer system, able to communicate with several outstations, is now offered by most control manufacturers as an alternative to a single, micro-computer based, central control system. A micro-processor at each out-station can then control directly all the actuators within its mandate by means of digital impulses (DDC) and a wide range of sensors provides the information necessary for control and monitoring. Comprehensive ranges of start-up, optimising and control functions are usually available within an out-station and changes of operational programmes or settings are possible manually at the out-station, or at the central micro-computer. The latter offers visual displays of status throughout the whole installation, supported by printers, alarms and keyboard facilities for the manual introduction of desired changes.

Much of the experience gained from the installation of energy management systems has been from refurbishment, and well-publicised case histories demonstrate their value in use (28). Although there is no doubt that the inclusion of such systems will be considered for all future, large or complex installations, designers must become more involved in their proper application, if the current difficulties being experienced in certain instances are to be avoided. Some of the application factors which should be considered are:

1 There is a need to deal efficiently and promptly with emergencies in large and complicated HVAC systems having the minimum of supervision. This, in itself, may be sufficient justification for a management system.
2 The benefit of potential savings in energy is often cited as a reason for the introduction of an energy management system but it should be noted that similar savings may be achieved with less capital expenditure, by using independent energy-saving packages.
3 A system may be created which is too complex for the operating staff to understand and manage. An outstanding problem here is the recruitment of people of the right quality to run the installation.
4 Compatibility between one computer system and any other offered by a different manufacturer is very desirable. Computer compatibility is essential if the performance of one building is to be compared with another or with a target (29). Early systems economised on memory usage but the reducing costs of micro-chip memories are resolving this difficulty.

While recognising the value of energy management systems, it is important to note that problems of application, installation and handover persist. Initially there was much difficulty and there were many failures because of the interference picked up from electromagnetic and radio-frequency sources but modern techniques can correct this. Problems of timing and setting to work, currently experienced, are being solved. For example, individual control loops can often be commissioned at any time and the supervisory functions of a management system added later, when convenient.

It must be understood that, with modern energy management systems, not only must there be a suitable clean area for the installation but the rest of the relevant HVAC systems must be completed and ready to operate before the management system itself can be set to work. The long period required for commissioning controls and proving software programs overlaps the periods normally allocated for setting to work and commissioning generally. If this is not recognised and an allowance made accordingly when planning the contract period, there will be uneconomic working for all concerned and the project will be completed late. Experience shows that several months must be allowed for proving these installations.

A further problem is the gap of knowledge between the designers of energy-management systems and the engineers in the field who commission them. The designers understand what the systems are able to do but the commissioning engineers may not always comprehend this sufficiently to implement their intentions.

Nevertheless, it is to be expected that energy-management systems will continue to be installed because of the overall benefits possible and that satisfactory solutions will be found to the contractual and commissioning problems that sometimes occur.

Natural cooling for air-water systems

Natural cooling is easily obtained with all-air systems in winter, when the outside air temperature is low enough, but a little more ingenuity has been necessary with air-water systems (induction or fan coil). The use of sprayed cooler coils to provide secondary chilled water at 11° to 13°C in winter has virtually ceased, for hygienic reasons, in commercial installations. In their place the closed-cooling tower or, perhaps better, the open cooling tower with a plate-heat exchanger, has been able to satisfy system needs for chilled water without having to run the refrigeration plant through the colder portion of the year.

More recently, the open cooling tower has been used directly and more effectively by the introduction of a proprietary strainer and a simple system of corrosion monitoring. Comparatively clean cooling water is circulated from the open tower through the secondary chilled water system. The penalty of two or three degrees, lost in the heat transfer processes at a closed tower or in a plate heat exchanger, disappears when an open tower and proprietary strainer is used. Natural cooling can be obtained for a longer period of the year and the saving in energy at the refrigeration plant is greater. The benefit is seen to increase the further north one goes in the UK.

The system has been extensively used for some time in the United States but has only been adopted during the last three or four years in the UK. Results in energy savings are so far encouraging.

Legionnaires' disease, humidifier fever and the sick building syndrome

Identified in 1976 (30) Legionella pneumophilla is difficult to detect, although it is now known to have existed way back (31,32). Probably present in 90% of cooling towers, it prefers a temperature range of 20° to 50°C and is common in HWS systems.

Air conditioning rightly uses cooling towers to cool water from 32°C to 27°C for refrigeration condensers in summer and HWS, in recent years, has tended to be stored at nearer to 50°C than the traditional 65°C, to conserve energy. With intermittent draw-off and piping heat loss, there may be many parts of an HWS system where the bacteria can thrive. It appears that HWS systems may be more to blame than cooling towers for the outbreaks that have occurred. The disease is acquired by inhaling a water droplet and not by ingestion. Nor can it spread from person to person. At the time of writing it seems that all cases of death from Legionnaires' disease have been in hospitals or hotels. There would seem to have been no recorded deaths from it in office blocks or other buildings.

Safety standards must be improved by proper design, installation and maintenance. Water temperatures in towers can probably be reduced to less than 20°C, for more of the year by the introduction of a secondary pumped supply (originally intended to prevent plumbing). Diligent maintenance with painstaking cleaning and the use of the correct biocide is essential for cooling towers.

Biocides cannot be used in HWS systems because of the danger to people but storage should be at above 50°C with a nightly lift to 65°C for one hour. Systems must be designed to avoid sludge pockets and dead legs where the bacteria can breed.

Humidifier fever is much less serious and is the result of an allergen that produces a transient, mild form of influenza in about 3% of the population of a typical office (33). If acquired at work on Monday, immunity develops in the early part of the week but is lost over the week end, the cycle repeating. The organisms causing this live in recirculatory spray water humidifiers in the range 5°C to 35°C and are difficult to identify. It is to be noted that since the maximum wetbulb in the UK is about 20°C and because most spray humidifiers work with water between 10°C and 15°C, Legionella pneumophilla is unlikely to breed in humidifiers.

Biocides should not be used for humidifiers of this type but proper cleaning and maintenance, with an automatic nightly draining, flushing and re-filling, when the air conditioning system is off, should deal with the risk of humidifier fever. It is prudent, however, to omit spray humidifiers entirely and to use dry steam injection when humidification is essential.

Capital and maintenance costs will rise slightly, as a result of these precautionary measures, but not by much. The price for health and safety is very small.

Building sickness seems to comprise (33): humidifier fever, occupational asthma (an allergy arising from spray humidifiers in about 8% of the population), rhinitis and eye irritation (not related to humidifiers but to mechanical ventilation), headaches and lethargy (not connected with air conditioning but apparently commoner in buildings with sealed windows).

People spend about 90% of their time in buildings having environments contaminated (34,35) with: bacteria, bio-effluents, volatile organic compounds from aerosol sprays, particulates, combustion products from tobacco smoking, radon gas, formaldehyde, viruses, fungus spores, the faeces of house mites and other pollutants. Many of these produce an allergic response in humans. The extensive use of plastics and glues in modern furnishings makes a significant contribution.

It is surprising, therefore, that a study (36) of conditions in two adjoining office blocks, one air conditioned and the other merely heated and naturally ventilated, showed levels of formaldehyde and carbon monoxide well below threshold limit values. Dry-bulb temperature, globe temperature, air velocity, relative humidity, moisture content, and ion concentrations were well within the accepted comfort zone. Nevertheless, there were many complaints of the work-related symptoms for building sickness, with more of them in the air conditioned building. However, it was not possible to identify any contribution made by the system: 100% fresh air was used, the spray humidifier was clean and conditions were comfortable. Psychological causes were considered but ruled out. Negative ion concentration was not significant.

This is not an isolated case, nor is it confined to this country. Research is continuing but has not yet solved the problem (37). The building services engineer must adopt the best possible standards for design, installation and, above all, maintenance to ensure air conditioning systems are beyond reproach.

The other side of the coin should not be ignored. A view is held in the United States that air conditioning, by proper filtration, can give rapid relief from certain allergic symptoms (37). Air conditioning for operating suites has achieved dramatic improvements in the control of bacterial infection. Our electronic society is only possible because of the ultra-clean air distribution provided by air conditioning in the manufacture of micro-electronic circuits and the successes in pharmaceutical research and production are entirely dependent on the sterile conditions made possible by air conditioning.

Trade union attitudes

In having a policy of seeking improved working conditions for all employees, the TUC has views on air conditioning. The major features of their attitude are as follows:

1. Putting pressure on the Health and Safety Executive to improve standards for the provision of fresh air.
2. Asking the Executive to set improved environmental standards at work.
3. Requiring better design and maintenance in air conditioning systems to avoid health hazards.
4. Urging consultation with employers before new building or refurbishment.
5. Reducing the exposure of workpeople to contamination by improving mechanical ventilation.

Air conditioning is accepted as giving better working conditions, provided it is well designed, installed and maintained. In the work place union officials are more concerned with the defects of air conditioning systems than their advantages. Although most accept the environmental benefits of good installations, complaints are made about those not working properly. Poor maintenance is criticised and the fact that the building occupants do not understand the capabilities and limitations of systems. The commonest problem is when a building is modified but the system is not: the provision of computers and other heat-liberating equipment, plus the introduction of partitions, frequently spoils system performance.

It is surprising there has been no particular union pressure. The overwhelming impression is that peoples' view of an air conditioning system is the same as their attitude to any aspect of change. There is an anticipation of difficulty in living in an air conditioned building for the first time, perhaps because of the inhability to open windows. Equally, they may then complain of the absence of air conditioning if moved out of the building to one that is only heated and naturally ventilated.

References

1 *The Building Regulations 1985, Part L, Conservation of fuel and power,* HMSO.
2 *The Building Regulations 1976, Part FF, Conservation of fuel and power in buildings other than dwelling,* HMSO.
3 Firth R., 'Automation brings many benefits but cannot cure all office ills', *National Computer Weekly,* pp. 14–15, 29th August 1985.
4 "Orbit Launched", *Architects Journal,* Vol. 180, pp. 51–53, 20th April 1983.
5 Korman R., "Why British offices can't cut it in the computer age", *Facilities Design and Management,* pp. 63–65, January 1984.
6 *Environment conditions in 10 Aberdeen offices,* an internal report, Mobil Oil Ltd., 1985.
7 *The decentralisation of offices from central London,* JLW Research, June 1983.
8 Weatheritt L. and John O.N.L., *Office development and employment in Greater London 1967–76,* Greater London Council Research Memorandum 556, 1979.
9 *City of London Information Report — Offices,* Dept. of Architecture and Planning, City of London Corporation, 1982.
10 Crawford P., Fothergill S. and Monk S., *The effects of business rates on the location of employment,* University of Cambridge, January 1985.
11 *Office Rental and Rates 1973 to 1975,* Debenham, Tewson and Chinnocks, May 1985.
12 Designs on the new consumer, *Marketing,* pp. 30–31, 24th October 1985.
13 *Research report: shopping in Bristol,* City of Bristol Planning Department, November 1983.
14 Something in store for building business, *Contract Journal,* p. 40, 26th July 1984.
15 Sainsbury's Annual Report and Accounts, 1985.
16 *Superstores and shopping centres,* Financial Times Survey, 23rd July 1985.
17 Success in store, *Building Services,* pp. 71–72, October 1983.
18 Riley R.L., Historical background in airborne infections, Ed. B.B. Kundsiu, New York, *Annals of the New York Academy of Sciences,* 353: 3–9, 1980.
19 La Force F.M., Airborne infections and modern building technology, Indoor Air, Vol. 1, *Recent Advances in the Health Sciences and Technology,* pp. 109–127, August 1984, Swedish Council for Building Research, Stockholm, Sweden, 1984.
20 Health facilities, *ASHRAE Handbook 1982 Applications,* Chapter 7, pp. 7.5–7.11.
21 Ellis J., Beyond the Nucleus, *CIBS Journal,* Vol. 4, No. 6, pp. 32–34, June 1982.
22 *DHSS Low Energy Hospital Study,* DHSS Publications, 1984.
23 *DD80 Laboratory fume cupboards, Part 1: 1982, Safety requirements and performance testing, Part 2: 1982, Recemmendations for information to be exchanged between purchaser, vendor and installer and recommendations for installation, Part 3: 1982, Recommendations for selection, use and maintenance,* British Standards Institution.
24 Government White Paper: *Scientific procedures on living animals — Cmnds 8883,* May 1983, HMSO.
25 *Building services for airports,* CIBSE Symposium, Gatwick, 6–7 November 1985.
26 *Computer based energy management in buildings,* ABBA Consultants, Industrial Survey Report, 1983.
27 *Centralised monitoring and control of building services,* CIBS Symposium, 1979.

28 *Energy Management Workbook*, Energy Publications, Cambridge Information and Research Services Ltd.

29 *Building Energy Code, Part 1 Guidance towards energy conserving design of buildings and services*, CIBS, 1977, *Part 2 "Calculation of energy demands and targets for the design of new buildings and services, Section (a) Heated and naturally ventilated buildings"*, CIBS 1981, *Part 3 Guidance towards energy conserving operation of buildings and services*, CIBS 1979, *Part 4 Measurement of energy consumption and comparison with targets for existing buildings and services*, CIBS 1982.

30 Cordes L.G. and Fraser D.W., *Legionellosis, Legionnaires' disease, Pontiac fever*, Medical Clinics of North America, Vol. 64, No. 3, p. 395, 1980.

31 *Facts and theories on Legionnaires' disease*, Industrial Water Society Conference, London, May 1981.

32 Wozniack S.J. *Legionnaires' disease — the implications for design and use of hot water systems*, Information Paper, Building Research Establishment, 1984.

33 Robertson A.S. and Sherwood Burge P., Building Sickness, *The Practitioner*, Vol. 229, pp. 531–534, June 1985.

34 Spengler J. Hallowell D., Moschandreas D. and Fanger O., *Environmental International (Special Issue): Indoor Air Pollution*, Pergamon Press Ltd., Oxford, 1982.

35 *Indoor Pollutants*, National Research Council, National Academy Press, Washington DC, 1981.

36 Robertson A.S. *et al*, Comparison of health problems related to work and environmental measurements in two office blocks with different ventilation systems, *British Medical Journal, Vol. 29*, pp. 373–376, 10th August 1985.

37 Physiological principles for comfort and health, *ASHRAE Handbook 1985 Fundamentals*, Chapter 8, pp. 8.1–8.32.

2 Building Envelope Design and Air Conditioning

Richard G. Saxon

Introduction

Air conditioning is used in the UK for some industrial process environments, for retail space generally, for deep plan elements of hospitals, and for office buildings. In all but offices the conditioned space either does not reach the envelope, or is separated from the outside world by a windowless wall or roof. Only in office buildings do we try to combine extensive window-walls with artificially controlled climate. Office-type space is becoming the principal wealth generating context as we move into the information society. Therefore, successful design of offices is of fundamental importance.

As to the use of exemplars as an alternative to methodology, several studies of the way designers relate to building science and research output suggest that they cannot absorb abstracted knowledge (1). They are, however, enthusiastic students of the work of others and seek relevant preconceptions to start the design process. Concrete examples of theory in action, presented or visited, lead to emulation, even if the theoretical principles remain only half digested. It should also be remembered that architects think holistically as well as empirically. Their search for solutions touches all influences on the envelope. Air conditioning will be one of them, but often a minor influence.

Air conditioning and offices

It is only possible to consider the envelope of a building type when the function within is clearly understood. I shall, therefore, work outwards from a discussion of the nature of the office building today to consideration of the envelope.

Twenty years ago we were being sold the concept of IED, Integrated Environmental Design. Air conditioning was put forward as a device which determined the ideal form of office buildings.

Deep-plan, open-plan space, lit to 1,000 lux and with vision-slit windows in insulated walls was portrayed as cost effective compared to naturally-lit and shallow-planned space. Burolandschaft came as part of the package. Heat-of-light, often greater than 50 W/m^2, provided the required warmth, and air conditioning was the method which enabled it to be used. This logic has been progressively undermined by changes in the perceived needs of office workers. The need for air conditioning now rests on quite a different foundation. Assuming that the building does not need to be sealed against noise or dirt, and is not overglazed, or over tall, air conditioning may still be required to dissipate the machine heat of the new office technology. As very many office buildings *do* have to cope with noise, dirt, height, and unshaded glass then air conditioning is needed for those reasons also. Plan depth and artificial lighting heat no longer figure as functional determinants in most situations.

Most crucially of all, air conditioning is now a market-led feature. Rental levels in the South East of England are based on definitions of prime and secondary space. Prime space is air-conditioned, just as it has good toilets and parking. Functional need is submerged within perceived prestige. Prime space in the provinces may not always be air conditioned, but agents will usually tell you that it is capable of being fully air conditioned if the tenant requires it. That requirement is only rarely based on technical performance.

How offices function now

Office building design is driven today by major trends in the nature of office work and occupier organisations. The clerical, low-tech factories of the sixties and seventies are being superseded by the information-technology based office, occupied by an organisation in the knowledge business. There are more senior people and fewer junior ones, and more concentrated, isolated working interspersed with more business and social gatherings.

Understanding of how office space works for organisations has greatly expanded, and several types of space are now wanted for different types of firm. A range of space plan depths help to accommodate diverse departmental needs, but most people will be based in the outer five metres of a floor plan. Internal space is increasingly used for shared IT workspaces, conference, machine and storage areas. Cellular offices are making a comeback as average status rises and concentrated individual work replaces group interaction. The divisive nature of multi-level offices is now recognised, and organisations are seeking larger floor plates to minimise divisions by level.

Working with information technology means working with continuous change. Three years is the obsolescence cycle of office machines, or twenty generations in the structural life of an office building. Designers and building owners are now becoming aware that services installations both wear out faster than building shells, and are made obsolete by user changes. Design of the permanent (40–60 year) shell of the building is becoming progressively decoupled from design of the 10–20 year service and scenery fit-out of the building. The envelope must live with an uncertain future as to what it envelopes. It must deal with the constraints, and not be locked into transient interior concepts.

The present technical requirements of premium office space can be summed up as: 300–500 lux ambient, plus task light, the ambient to be from ultra low brightness ceiling fittings or uplighters; views out but very low window brightness; air-conditioning to maintain comfort standards under the impact of 15–20 W/m^2 lighting load; 10–60 W/m^2 machine load (but typically 20 W/m^2), with static controlled; space capable of full cellular subdivision; cable management able to handle power, telecom and data, or combined digital/LAN systems, interfacing with suitable furniture (2).

Typical current answers to this range from the semi-industrial 'everything in the ceiling' approach, through combined cavity floor and suspended ceiling systems, to radical 'everything in the floor' approaches. Floor to floor height is generally pushed up compared to the specification typical before 1980. There is also the 'super office', as a standard setter, the City of London financial centre with a need for deep-plan space, column free, with further enhanced storey heights, floor loadings and heat dissipation capability.

The range of technical solutions to the matter of air supply and removal, cable management, lighting supply and furniture ergonomics is widening. And with this diversity the concept of a predictable interior fit-out, which can be integrated with the design of the envelope, is disappearing. An office building completed tomorrow would see its tenants climb the office technology ladder progressively, and at different speeds. Initially they may not need artificial cooling, or underfloor service facilities. Eventually they probably will, and the building will be modified step by step, space by space.

How offices are marketed

The envelope of an office building provides a lot of its market image, as well a packaging the usable space. Envelope design decisions for market reasons are more powerful than those for air conditioning cost control and should, therefore be considered here.

The envelope's inner face defines the rentable area of the building. This is a vital fact for environmental design. In the USA floor space is measured to the back of the glass, and this has been a prime cause of the triumph of mirror glass. Minimal wall thickness, right down to the carpet, with the outer face actually on the property line or zoning envelope, is economically essential. In the UK floor space is synonymous with carpet area, which allows more scope for wall thickness, though not a lot. With the increased stress on floor plate size developers can begrudge solutions with any depth of structure or external shading around the perimeter. Wily agents, negotiating rental levels, are increasingly alive to the question of the usefulness of the outermost 1–2 metres of space. Floor level air grilles may be disputed as rentable area. A glass wall with poor insulation or shade value makes the outermost space uncomfortable for anyone to sit in. As the outermost band can represent a very significant percentage of the lettable area, these are major value issues. What we have not yet seen here is the US trend to multiplying the number of 'corners' in the plan. Corner offices are much sought after as a sign of high status, and command a higher rental ahead of their extra cost. Hence the serrated plans increasingly common in American office buildings.

An important influence is the increase in multiple tenancy in office buildings. Single tenancy is always the goal of investing institutions, as it simplifies administration and increases lettable area. Multiple tenancy is an inexorable trend, however, as buildings get bigger and tenants' operating units get smaller under the influence of IT. Floor by floor or suite-by-suite servicing is the logical outcome. Even when buildings are let as a whole, or built for corporate use, subdivisibility of the building is a powerful demand today, to enable the prime leaseholder or owner to reduce space use easily if need be, and to give highest investment value to the whole property.

A major trend now gathering pace in the City of London is the adoption of the American approach to letting 'raw' space rather than finished. In Britain it has been traditional to pre-finish office interiors to a high standard, then offer them to tenants. In the US the bare shell is marketed, with a customised fit-out negotiated around a notional standard specification. As the actual needs of tenants have become more diverse, British tenants have found themselves tearing out the landlord's interior and servicing and replacing them, at great cost. The idea of renting just the shell and core of the building is both tax efficient and highly practical, but requires vision and skill from the tenant which only major space users already have. It will spread as a concept, however, as the development and design industries provide the service to help tenants take full advantage.

Thus the practical decoupling of the envelope from the interior is being followed by legal decoupling. Tenants will be increasingly responsible for their air conditioning, suite by suite, and for their ceiling and floor distribution concepts. Basic building systems which allow for incremental servicing of each suite are predicted in the Obit study (3). One of the ramifications of this is that individual tenancies will need external wall locations where air-handling plant can be added, or where cooling plant can breathe. Rooftop and terrace sites for plant and telecommunication gear, to avoid use of expensive rental area, will be more desirable. Owner occupied space will be similarly serviced as different parts of a building come to house departments at different levels of IT use. Lloyds of London provides a caricature of the concept, with a pure box of usable area to which all services are externally attached. These outer elements have grown extraordinarily in the five years the project has been in progress. They will continue to grow and change as internal needs shift. None of this plumbing need be visible, of course, and a few US and UK buildings are similarly sophisticated behind a conventional or even historical facade.

It is ironic that the trend away from fully glazed walls to more solid, massive and hence shaded forms has more to do with official conservation policy in cities, the post-modern preferences of some architects and conservatism amongst clients than any concern for operating costs. It too is a market factor affecting office envelopes. So is the emergence of the atrium, to which I shall return. It is included essentially for marketing reasons, with any performance effect on air conditioning being quite incidental.

Relative costs are important to the market view of air conditioning. Extra annual operating and maintenance costs for systems average 74p per square foot, out of a total difference of £1.27 per square foot per annum between air conditioned and non-air conditioned property (4) (5). The rest of the difference is due to the generally higher maintenance and service level of prime property. Air conditioning as a marketing necessity is a creature of prime sites where rent, rates and service charges together range from £25 to £50 per foot per annum. In this context the economics of air system design are marginally important. Outside Greater London, where space occupancy costs fall to between £8 and £12 per foot, the importance of system operating costs is much more, but of course air conditioning here is the exception rather than the rule. Hence there is a cruel market logic inhibiting demand for economic design: where air conditioning is required by the market, its operating costs are inconsequential.

The envelope which results

Now we are able to consider the sort of envelope design which will answer the functional and market forces of today. Can passive solar energy help in any way? Several studies suggest that it can, but principally through saving lighting energy (6) (7). But we have already seen that bright windows are incompatible with VDU working. Only indirect daylighting is, therefore, likely to be acceptable, with the majority of glazing used for view only and of reduced brightness. Division of window areas into lighting clerestory and view areas follows logically, and with it ideas on the location of shading. An internal light shelf at about 2m from the floor, baffling view of the clerestory, can be extended outside to provide shade to the vision window.

Effective solar shading is important to refrigeration economy, but also to comfort in the outer 2m of floor space. Such shading, with preserved view, is only achieved simply on South and North elevations and it is, therefore, likely that the architect will try to orient most space in these directions. This orientation also allows use of passive solar heating or cooling. As an office with a typical level of IT installed is unlikely to want any general supplementary heating, the contribution from passive solar heat can best be used to overcome cold radiation in the critical outer 2m of floorspace. Radiator systems or air inputs doing this job are space consuming as well as energy consuming. Using double skin external walls as at Leslie & Godwin, or atria, allows the inner skin to be warmed easily and to retain its heat.

Another intelligent recent application is at Finsbury Avenue, where the window wall mullions are made into a plumbing system and warmed, thus directly warming the glass and avoiding space consuming perimeter heating. The use of atria is accelerating in the UK, and the concept does free the inward facing accommodation from the need for much performance from a partition overlooking an atrium. Indeed the wall can be, and is occasionally, omitted entirely. This, however, implies full comfort standards in the atrium, and pushes the envelope design issue out to the atrium boundary. The great majority of UK atrium buildings have sealed windows to surrounding accommodation, for fire safety and acoustic reasons, and are air conditioned at comparatively low cost. There have been some designs which use the atrium as a supply duct routeway (Drummond Gate) and as a return air plenum (mostly in North America). Input and outlet grilles follow. The atrium is a 'soft' routeway for changing service installations, and will probably figure further in designs for flexibility. (8)

I have mentioned earlier the logic of suite-by-suite air handling, rather than central a/c plant. Smaller, but still reliable plant is now with us which breathes directly from outside, with plant accommodated in minor rooms, or even in service void depths alone. The ability to add to or move these installations is welcome, again with implications for external wall detail.

What of the mass requirements of external walls? High mass appears to be an optional feature. As more steel framed, decked and metal walled buildings emerge they use the rapid thermal response which results to warm up quickly for their intermittent occupation. Good external shading is essential to economy, and is increasingly seen on such buildings, in the form of outrigger shades or mirror glass. Heavy walling, in masonry or precast concrete, is usually decided upon for reasons of aesthetic context or taste rather than pure function. The advantage of thermal storage capacity in the envelope is slight but real. When weight is not a problem and windows are to be smaller than in curtain wall solutions it is possible to achieve satisfactory shading factors quite easily. Once sun is admitted, air warmed by it will need to warm the outer wall, and ideally also the floor slab, before it heats up too much itself. Heavy floors are not directly open to the solar admittance at all these days, as carpet is universal and elevated floors are more common. However, the soffit is capable of exposure to warm room air, either sun warmed or heated by office lighting and equipment. An exposed floor slab is a cool radiator, and slows room temperature build up by 1–2 °C, effectively saving refrigeration load. Even when overhead acoustic control and lighting are used, it is possible to keep the slab soffit exposed to room air, or to use it to cool plenum air supply or return. With elevated floors it is, of course, possible to use the top of the slab as a thermal reservoir to cool supply or return air. Some designers have considered using hollow reinforced concrete floors to duct air, thus cooling slab surfaces. Such installations can be pre-cooled with night air, reducing peak loads further.

Operable windows need not disappear entirely in well sited air-conditioned offices. Sealed buildings can use mid-season air directly for free cooling, but at some cost in running the whole system. Shallow plan buildings can use mid-season air directly through operable windows, enhancing occupant control. Windows would be fully closed in winter and summer when heat or cooling are needed. Only at the top of the IT ladder is it probable that there will not be any scope for outside air use direct. Operable windows to atria are possible also, provided they are powered, and can be shut automatically along with other countermeasures in the event of fire alarm.

One final and unusual air-conditioning influence on envelope design is the increasing variability of floor and ceiling levels. Until the opening up of underfloor servicing options one very predictable element in design was finished floor level, and all the related levels of window sills, transoms, doorheads, lift, toilet and stair provisions. Over the life of an office building of today it is possible that services may be both overhead and underfloor, or entirely one or the other. Finished floor levels could vary by 600mm as a result. Dcmountable window wall panels, able to respond to these shifts, could be one answer. Slab to slab glass, with variable applied shading, insulants or other controls, will be another. Office block cores will respond by being more detached from the usable area, with transition corridors which can be ramped up or down.

An exemplar — refuge assurance HQ

This is a corporate headquarters building by BDP. It is out-of-town, in the Manchester suburb of Wilmslow, and is in mature parkland. Because the site is quiet and clean and the building low-rise, the only reason to air condition was to cope with machine heat.

The building follows the current imperatives by having large floor sizes, but shallow plan space: 4,000 square metres per office floor, with 12–15 metre wide space. The plan is made up of three repeated 'animals' of space, with their limbs connected to enclose two courts. Each animal can be sublet and others can be added if need be. Each is serviced by a roofspace plant room. Air, power, data and telecom systems all flow in a 600mm underfloor cavity, with the soffit of each flat floor slab exposed as a heat sink and light reflector. Uplighting and indirect daylighting are teamed together.

The external walls are masonry with double glazed operable windows well inset. The plan form gives shade to many windows, but exposed ones have blinds between the panes of each window. The roof is massive, with plant below it and a full floor slab above the topmost office giving the same conditions as below. Solar impact is thus minimal.

At occupation the machine heat load outside computer rooms proper is not going to be heavy enough to require cooling on more than the warmer summer days. The operable windows will, therefore, generally be used at mid-season. In winter and high summer windows will be closed with air supplied heated or chilled. As a further mid-season economy, when outside temperatures permit, chilled water can be provided by evaporative cooling only, by-passing the compressor. As machine heat builds up so periods with windows closed and full air conditioning will rise. Air supply is by fan coil units buried in the floor void, with return air in the floor plenum. Air can also be supplied and removed through high-load individual work-stations when this is justified. The general weight of the envelope saves cooling power by delaying the build up of solar heat, increasing the time when natural ventilation can be comfortably used.

Conclusions

This chapter has attempted to show that an architect sees the building envelope from many different angles at once. The effect of air conditioning in the form, massing, fenestration, and materials used is relatively slight. Firstly because air conditioning costs are a modest increment on those properties where its use is required. Secondly because design decisions to minimise purchased energy cost are much the same without or with air conditioning. But the overriding form giver of air conditioned buildings is the intended end use. Office use, the dominant one, has its own imperatives which lead to the prime decisions. Design for efficient air conditioning is then a matter of detail in meeting the strategic needs of the office market.

References

1 Hillier, Musgrave, O'Sullivan, Knowledge & Design, *Environmental Design, Research and Practice,* UCLA, 1972.
2 Vincent G. and Peacock J., *The Automated Building,* Architectural Press, 1985.
3 DEGW and EOSYS, *The Orbit Study,* 1983.
4 Jones, Lang, Wooton., *Office Service Charges,* Occasional Paper, Winter, 1984–5.
5 Henley Centre, *The Scope Report, a study of the cost of office premises in England,* 1985.
6 Building Design Partnership, *Low Rise Office Block, report on design intent,* ETSU document no. DSC/P–1, 1985.
7 Duncan I. (Building Design Partnership) and Hawkes D. (The Martin Centre, University of Cambridge), *Passive Solar Design In Non-Domestic Buildings, a report to the Energy Technology Support Unit,* Harwell, June 1983, Ref. No. ETSU-S-1110.
8 Saxon R.G., *Atrium Buildings, development and design,* Architectural Press, 1983, 1986.

3 Air Conditioning and Energy — the CIBSE Building Energy Code, Part 2(c)

W.P. Jones

Introduction

The principles underlying the concept of energy targets for the design of systems for new buildings has already been explained and the case well argued (1). It is worthwhile, nevertheless, to say a few brief words on the conception of the Building Energy Code in order to explain the rather long gestation period of Part 2(c), which deals with air conditioned buildings.

The original intention of the Working Party was to produce a single Building Energy Code, covering all aspects of energy demand and consumption and proposing comprehensive targets. The magnitude of this task became apparent very rapidly and the proposed Code was split into four sections, parts 1 and 2 dealing with the design of new buildings and parts 3 and 4 with the performance of existing buildings and systems. Parts 1, 2a (3), 3(4) and 4(5) have already been published.

The reason why Part 2 has been sub-divided into three further parts, (a) dealing with heated and naturally ventilated buildings, (b) with mechanically ventilated buildings and (c) with air conditioned buildings is because of the problems involved and the time-consuming nature of attempts at their solution. The establishment of a method for estimating thermal and electrical demands for heated and naturally ventilated problems was achieved comparatively easily although difficulties arose in allowing for the benefits from daylight and solar gains. A major matter of concern was in determining a method for assessing thermal and electrical targets and it is probably only because one other member of the Working Party was able to propose a dimensionless quantity, termed the Building Envelope Number (abbreviation B), that a satisfactory result was possible.

Section 2(b), which should be published towards the end of 1987, has relied upon Part 2(a) to determine thermal demand, a mechanical ventilation rate being substituted for a natural rate. Targets are again expressed in terms of the Building Envelope Number, suitably modified to take account of the mechanical ventilation rate. The work of the Building Research Establishment (6) on over-heating in offices was relied upon when deciding on acceptable mechanical ventilation rates for determining targets.

The work on Part 2(c), dealing with air conditioned buildings, has proceeded in parallel with that on Part 2(b). That it is not yet ready for publication arises from the formidable nature of the problems that air conditioning systems pose, when trying to establish methods for assessing demands and proposing credible targets. These difficulties stem from the complicated structure of the cooling load, the wide variety of systems possible, the range of uses to which they may be applied and the scope open to architects when designing buildings themselves. What follows may, therefore, be regarded as an interim report on progress with a prognosis on developments and, hopefully, successs to come.

Method adopted

The electrical power required by the services in an air conditioned building is the sum of the following components (due attention being paid to diversification): electric lighting, the plant associated with the heating and hot water systems, the refrigeration plant and its ancillaries and the fans, pumps and compressors associated with the rest of the air conditioning and ventilation systems. Power consumed by machines related to the industrial or commercial activities in the building is excluded as not being part of the building services.

The needs of the lighting, heating and HWS systems are dealt with in Part 2(a). Part 2(c) therefore concerns itself with the air conditioning power requirements. As was the case with Part 2(a) the annual average rate of energy use is obtained by the multiplication of the maximum absorbed power with a suitable load utilisation factor and it is in the determination of such utilisation factors, for items of air conditioning plant, that the major thrust of Part 2(c) lies.

The power consumed by items of plant such as fans or pumps that run at a constant rate is easy to assess. Refrigeration plant is much more difficult to deal with because it runs at partial load for the majority of its working life and the same can be said of variable air volume systems although the problem they present is resolved more easily. In view of the importance of the contribution made by the refrigeration plant it was decided to formalise the calculation of the duty of the cooling system by prescribing a method for the establishment of a standard refrigeration load (Q_{rs}), incorporating standard values for certain items, where this seemed appropriate, as an aid to simplification.

Having determined maximum loads for each item of equipment in the air conditioning system, the individual mean load is calculated for each month of the year using suitable (standard) diversity factors and taking account of variations in the weather. Knowing such mean monthly loads it is then possible to calculate a load utilisation factor for each piece of plant and so to evaluate a mean power consumption for the year.

If a method can be established to quote a target then it becomes possible to compare an electrical demand with an electrical target, for a particular system in a particular building. Although different systems of air conditioning will necessarily give rise to different electrical demands it is hoped that the targets to be quoted will not be expressed in terms of particular systems but will be related to applications in different types of building (offices, hotels etc). In this way the ingenuity of the designer to meet a target will be encouraged and bias towards a particular system will be avoided. One aspect of design that will have a significant influence on the comparison between demands and targets is the form of automatic control adopted and similar considerations apply to thermal demands and targets.

Whether these intentions can be fully realised remains to be seen. In particular it appears likely that proposing targets that will satisfy the variety of systems on the one hand, and the wide range of building applications on the other hand, is going to be difficult. There is the further question of the variations possible in building form. In what follows the attention given to some of these problems is explained in more detail.

Standard refrigeration load

Calculation method

The standard refrigeration load (Q_{rs}) must be calculated for the building and its air conditioning system according to a prescribed method. It is expressed by;

$$Q_{rs} = (Q_s + Q_{lh} + Q_{fa} + Q_d + Q_{fp} + Q_{rh} + Q_{ra})\,\phi_{pp}\ \text{kW}$$ **(equation 1)**

where:

Q_s = sensible heat gain, kW

Q_{lh} = latent heat gain, kW

Q_{fa} = fresh air load, kW

Q_{fp} = fan power, kW

Q_d = duct heat gain, kW

Q_{rh} = re-heat, kW

Q_{ra} = heat gain to return air, kW

The term ϕ_{pp} is a factor to cover the small contribution made by heat gain from pumps and pipes and it is specified that the standard load be calculated for a standard time, namely, 1500 h suntime in July.

Not all the elements have the same significance in equation 1 and simplifying assumptions are made in some of them for this reason. It is quite possible that, as the structure of the method is developed, it will be necessary to make more simplifying assumptions to reduce the quantity of calculation involved. Common sense will dictate this.

It is evident also that the load will be different for different systems. For example, high velocity systems will have a larger contribution for fan power (Q_{fp}) than will their low velocity counterparts. Furthermore, the same application in buildings of different design will result in different standard refrigeration loads dictated, for example, by different areas of glass, different types of solar shading or different building orientations. These possibilities are seen when the elements of the standard load are examined a little more closely.

Sensible heat gain

This is calculated for the air conditioned portion of the building as a whole, at 1500 h suntime in July, and is expressed by:

$$Q_s = (\Sigma Q_{tw} + \Sigma Q_{tr} + \Sigma Q_{tf} + \Sigma Q_{tg}) + Q_{si} + Q_{sg} + Q_{el} + Q_{sp}\ \text{kW}$$ **(equation 2)**

where:

ΣQ_{tw} = the sum of the transmission (air-to-air) heat gain through all the walls, in kW

ΣQ_{tr} = the sum of the transmission heat gain through all roofs, in kW

ΣQ_{tf} = the sum of the transmission heat gain through all floors exposed to the outside air, in kW

ΣQ_{tg} = the sum of the transmission heat gain through all the glazing, in kW

Q_{si} = the sensible heat gain by natural infiltration, in kW

Q_{sg} = the sensible cooling load on the air conditioning system arising from solar gain through glass, in kW

Q_{el} = the sensible heat gain from electric lights, in kW

Q_{sp} = the sensible heat gain from people, in kW

The CIBSE method of sol-air temperatures is used to calculate heat gain through walls, but, in the case of roofs, as a simplification, solar effects are ignored for buildings of more than four storeys. A standard allowance of 0.5 air changes per hour is adopted for calculating the sensible gain by infiltration and, in the case of the solar gain through glass, air point control is assumed. Standard diversity factors are tabulated for people and for electric lights according to the application and, where lights are concerned, according to the amount of glazing and the distance inwards from the line of the windows.

Where the value chosen by the designer for the outside air temperature at 1500 h suntime in July is different from the value given in CIBSE table A2.23, the sol-air and dry-bulb temperatures in the table are augmented by the difference, prior to use in the calculation of the sensible heat gain.

Latent heat gain

This poses few problems of calculation if the standard value of 0.5 air changes per hour is accepted for infiltration and is expressed by

$Q_{lh} = Q_{li} + Q_{lt}$ kW **(equation 3)**

where:

Q_{li} = the latent gain by infiltration, in kW

Q_{lp} = the latent gain from people, in kW

The outside and inside design conditions chosen by the designer of the air conditioning system in conjunction with the standard infiltration rate yield a value for Q_{li} quite easily. Standard diversity factors and population densities are provided and Q_{lp} is not difficult to calculate.

Fresh air load

Standard fresh air allowances are given in terms of the application and the population density. A standard specific volume is also quoted and the fresh air load is then determined by:

$$Q_{fa} = (x\dot{v}/0.8674)(h_{os} - h_{rs})\ \text{kW} \qquad \textbf{(equation 4)}$$

where:

Q_{fa} = fresh air load, in kW

x = the minimum fraction of fresh air, by mass

$\dot{v}$ = supply airflow rate, in m^3/s

h_{os} = the outside air specific enthalpy at 1500 h suntime in July in kJ/kg

h_{rs} = the room air specific enthalpy at 1500 h suntime in July in kJ/kg

Fan power

This is the consequence of the temperature rise suffered by the airstreams flowing through the supply and extract fans. The temperature rise is independent of the airflow rate and, if we make reasonable assumptions for fan efficiency, can be regarded as about 1 degree rise per kPa of fan total pressure. Standard temperature rises can then be specified for different systems. With such standard values the fan power contribution to the standard refrigeration load is determined from:

$$Q_{fp} = 1.22(\dot{v}\,\Delta t_s + (1 - x)\,\dot{v}\,\Delta t_e)\ \text{kW} \qquad \textbf{(equation 5)}$$

where:

Q_{fp} = fan power, in kW

$\dot{v}$ = supply airflow rate in m^3/s

Δt_s = standard temperature rise across the supply fan, in K

x = the minimum fraction of fresh air, by mass

Δt_e = standard temperature rise across the extract fan, in K

Equation 5 is a volumetric form, that can easily be derived, of the fundamental relationship between heat flow rate, mass flow rate, temperature rise and specific heat capacity.

Duct heat gain

This is probably seldom calculated for actual designs, unless a computer is used. It may not prove to be very significant and probably it will be possible to adopt a standard value or even ignore it completely. At the moment it is included and calculated by:

$Q_d = 0.5 \times 1.22\,\dot{v}\,\text{kW}$ **(equation 6)**

for high velocity systems and by

$Q_d = 1.0 \times 1.22\,\dot{v}\,\text{kW}$ **(equation 7)**

for low velocity systems

where:

Q_d = duct heat gain, in kW

$\dot{v}$ = supply airflow rate in m^3/s

Temperature rises of 0.5 K and 1.0 K are assumed for high and low velocity duct systems, respectively.

Re-heat load

At the time of writing this is posing a serious problem. Intrinsically, the calculation of re-heat load is easy and is given by

$Q_{rh} = 1.22\,\dot{v}\,(t_{ro} - t_{ri})\,\text{kW}$ **(equation 8)**

where:

Q_{rh} = re-heat load, in kW

$\dot{v}$ = supply airflow rate, in m^3/s

t_{ro} = temperature at re-heater outlet, in °C

t_{ri} = temperature at re-heater inlet in °C

Few types of air conditioning system have the same re-heat characteristics and this means that each system possibility should be considered separately. There appear to be about a dozen different systems that could be used for air conditioning over the range of commercial and industrial applications likely.

Although it may be possible to simplify this situation for the calculation of standard refrigeration loads, it would not be possible to do so when calculating the mean monthly loads where re-heat often plays a surprisingly significant role.

Heat gain to return air

The importance of this depends on the extent to which recirculated air is used and whether extract ventilated luminaires are adopted. Its value is given by:

$$Q_{ra} = 0.4\, f_l\, W_l\, (1 - x)/y \text{ kW} \qquad \textbf{(equation 9)}$$

where:

Q_{ra} = return air load in kW

f_l = standard diversity factor for lights

W_l = the maximum power absorbed by the lights in kW

x = the minimum fraction of fresh air, by mass

y = the fraction of the supply air quantity, by mass, that is extracted through the luminaires.

Equation 9 is obtained from the standard assumption that 40% of the heat liberated by the lighting is absorbed by the extract airstream and appears as a load on the air handling plant, after allowing for the fact that the warmer air in the ceiling void will transmit some heat upwards through the floor slab and downwards through the suspended ceiling. Using the same notation as for equation 9 this is typified by

$$0.4\, f_l\, W_l = 1.22\, y\, \dot{v}\, (t_e - t_{rs}) \text{ kW} \qquad \textbf{(equation 10)}$$

where $\dot{v}$ is the airflow rate in m^3/s, t_e is the temperature of the extracted air in the ceiling void and t_{rs} is the room air temperature in the summer at 1500 h suntime in July.

Since we also know that the return air load is given by

$$Q_{ra} = 1.22\, (1 - x)\, \dot{v}\, (t_e - t_{rs}) \text{ kW} \qquad \textbf{(equation 11)}$$

with the same notation, it is possible to re-express equation 11 in terms of $(t_e - t_{rs})$ and substitute in equation 10 to yield the simpler form of equation 9.

Contribution of pumps and pipes

This is expressed as a factor that multiplies the rest of the standard refrigeration load. Standard values of ϕ_{pp} for all-water, air-water and all-air systems are given. The percentages involved are quite small, of the order of 1%, and it is likely that this element will be dropped, to simplify the arthmetic for such a complicated subject.

Influences on refrigeration demand

Partial refrigeration load

The refrigeration plant does not run at a constant duty throughout the year because the values of the component elements in the load, expressed in standard form be equation 1, vary individually and collectively. Furthermore, where the outside air is cold enough it is possible to switch off the refrigeration plant and obtain so-called free cooling, supposing that the system has been designed to exploit this situation. The type of system and the way it is controlled also exert an influence on the benefit from free cooling.

It follows that having calculated a standard refrigeration load it is necessary to determine the partial duties at which the plant would be likely to run throughout the year. For this purpose meteorological data from the CIBSE Example Weather Year (1 October 1964 to 30 September 1965 at Kew) has been used where appropriate. (Not all the components of the standard refrigeration load are affected by the weather.)

Each component of the standard load is examined in turn, during the periods that the refrigeration plant will run, throughout the year. To do this, it is first necessary to establish the running times for different types of system and modes of control.

Operational hours of refrigeration plant

For the case of all-air systems a standard outside dry-bulb temperature of

10°C has been adopted, it being assumed that when the outside air temperature is less the refrigeration plant can be switched off and an air temperature of 10°C obtained by the thermostatic mixing of outside and recirculated air. In analysing the data from the Example Weather Year account is taken of the operating period of the air conditioning system. Tabulated data is provided of mean values of dry-bulb temperature, enthalpy and moisture content for a working day of eight hours from 0900 h to 1700 h, due consideration being paid to the plant being off at week-ends and on eight statutory holidays. The changeover from GMT and BST and vice-versa has also been accounted for.

A similar analysis has been done for the case of systems using fixed proportions of fresh air which have, obviously, far less scope for free cooling.

To cover the case of air-water systems, where the free cooling option depends on there being a sufficiently low outside wet-bulb temperature, a standard outside wet-bulb of 5°C (screen) has been adopted.

Finally, the above analyses have been repeated for systems that run continuously, 24 hours a day, 7 days a week.

Mean monthly refrigeration loads

We can infer from the foregoing that the operating hours when the refrigeration plant must run can be calculated on a monthly basis, for different types of system. To determine the energy used for mechanical refrigeration (and so establish its demand) it is necessary to calculate the mean cooling duty, throughout each month of the Example Weather Year when the refrigeration plant must run.

Having determined the mean monthly load, its product with the operating time in that month will yield the refrigeration energy. When this is divided by the total in the year (8760 hours × 3600 seconds/hour = 31536000 seconds) the average annual refrigeration duty is obtained. Dividing this again by the standard refrigeration load gives a utilisation factor for the refrigeration plant.

It is evident that in determining the mean monthly refrigeration load, each element in equation 1 that contributes to the load must be individually calculated. In doing this for the mean solar gains through glass ($\bar{Q}_{sg}$) solar data from the Example Weather Year is used and similarly, where appropriate, other data for the other elements. Variations in the room state, with respect to temperature, moisture content and enthalpy are relevant to the calculation of transmission gains through the building envelope, the latent gains and fresh air loads. Accordingly it is assumed that room temperature and moisture content vary in a simple sinusoidal manner between their summer and winter design values to simulate the influence of the proportional bands in the automatic controls.

Mean monthly refrigeration power

To determine the electrical demand of a vapour compression refrigeration plant, an estimate must be made of its coefficient of performance (abbreviation COP) at full and partial load. For purposes of the Code (2) the efficiencies of all components in the refrigeration plant that use energy must be included. Similar considerations apply to vapour-absorption refrigeration plant which have a thermal as well as an electrical demand.

The thermodynamic aspect of the COP depends principally on the difference between the condensing and evaporating pressure and as a simplification it is assumed in this Code that evaporating pressure is controlled at a constant value. For the case of centrifugal compressors it is supposed that condensing pressure is allowed to fall, in a controlled manner, down to a safe, lower limit. On this basis it is reasonable to make the simplifying assumption that the power absorbed by the compressor motor is directly proportional to the cooling load. We can then write:

$$\bar{W}_{ref} = (\bar{Q}_{rm}/Q_{rs})\, W_{100}\ \text{kW} \qquad \textbf{(equation 12)}$$

for centrifugal machines, where:

$\bar{W}_{ref}$ = mean monthly electrical power absorbed in kW

$\bar{Q}_{rm}$ = mean monthly refrigeration load, in kW

$\bar{Q}_{rs}$ = standard refrigeration load, in kW

$\bar{W}_{100}$ = electrical power absorbed at the standard refrigeration load, in kW

With reciprocating machines the condensing pressure must be held constant. Other factors, such as friction, then exercise an increasing influence as the duty falls and the electrical power used by the compressor motor is not directly proportional to the cooling duty. To simplify the matter the following equation gives an approximate relationship between cooling duty and absorbed electrical power for reciprocating compressors over a capacity range, per machine, from 100% to 25%.

$$W_p = 100 - (100 - Q_p)\,5/8\% \qquad \textbf{(equation 13)}$$

where:

W_p = percentage of full-load electrical power

Q_p = percentage of the standard refrigeration load in any particular month.

If the mean monthly refrigeration load, $\overline{Q}_{rm}$, and the electrical power absorbed at the standard refrigeration load is known then the mean monthly electrical power, $\overline{W}_{ref}$, absorbed by the compressor motor, can be calculated and with the same notation we have:

$$\overline{W}_{ref} = W_p\, W_{100}/100\ \text{kW} \qquad \textbf{(equation 14)}$$

for reciprocating compressors.

The energy used by a refrigeration plant

Above it was seen that a utilisation factor could be established for each month that the refrigeration plant must run. Equations 12 or 14 will then yield the corresponding mean monthly electrical power for the refrigeration compressor(s). It is evident that the running hours in each month can be evaluated. The product of the running time and the mean electrical power then gives the mean electrical energy used by the refrigeration compressor(s) in each month and hence for the year as a whole.

Other plant items

Refrigeration plant auxiliaries

These are principally pumps for chilled and cooling water and cooling tower fans or air-cooled condenser fans. Most of the pumps run at a constant duty for a period equal to that of the refrigeration plant, although some pumps may run for longer periods to provide free cooling by various means. It is not difficult to calculate the electrical energy used through a year.

The fans on cooling towers or air-cooled condensers often run intermittently, under automatic control for part of the year and therefore pose more of a problem than pumps. However, because their energy consumption is not likely to be a very large fraction of the energy consumed by the whole system, simplifying assumptions can be made to give reasonable answers.

Supply and extract fans

The evaluation of the energy consumed by the fans in constant volume systems is easy to calculate.

With variable air volume systems the calculation is not quite so straightforward. However, since the air volume handled is directly proportional to the sensible heat gain, down to a minimum value dependent on the type of air supply terminal unit, mean monthly airflow rates can be calculated because mean monthly sensible heat gains have been calculated earlier. A practical problem arises in that very few variable air volume installations so far have been commissioned properly and it is possible that this poor practice may continue. It arises from a failure to control supply and extract fan capacities in a manner that tracks the turndown of the air distribution terminals. For the purpose of this Code it must be assumed that this practice will cease and that future systems will be properly controlled.

Present position and future development

Calculation of demand

The foregoing text describes the present situation and it will be seen that it relates entirely to calculating the electrical demand. The methods proposed have been provisionally tested for a simple office block having its major axis aligned N–S and treated by a two-pipe, non-changeover induction system. Tentative results are given in Table 3.1 which must contain a few inconsistencies, if examined closely, because of the extent of the computation involved and the provisional nature off the calculations. Refinement and possible simplification will help to resolve this.

The Table refers to one water-cooled centrifugal water chiller and a pair of these are used to deal with the hypothetical office block of 14,000m^2 floor area used as a model for the example. Two such machines consume twice the sum of the mean monthly values for energy used in Table 3.1, namely 2 × 716.3 = 1433 GJ per annum. At a calculated standard refrigeration load of 1519 kW each machine absorbs 169 kW of electrical power. Thus the energy used by two machines running at full load would

Table 3.1 *Mean monthly electrical energy used*

MONTH	J	F	M	A	M	J	J	A	S	O	N	D
% Q_{rs}	64	56	68	65	77	83	78	86	77	69	68	67
$\bar{W}_{ref}$ (kW)	108	95	115	110	130	140	132	145	130	117	115	113
Hours run	45	19	109	151	152	200	168	200	200	157	138	49
Energy used (GJ)	17.5	6.5	45.1	59.7	71.2	101.0	79.7	104.6	93.7	60.2	57.1	20.0

be 2 × 169 × 8760 × 3600/10⁶ = 10659 GJ and the utilisation factor for the refrigeration compressor motors is 1433/10659 = 0.134.

Making reasonable assumptions for the operation and efficiency of the associated pumps and fans we have the results shown in Table 3.2.

Table 3.2 *Mean annual electrical powers of refrigeration plant and auxiliaries*

PLANT ITEM	FULL LOAD POWER kW	UTILISATION FACTOR	MEAN ANNUAL POWER kW
Compressors	338	0.134	45.3
Cooling water pumps	30	0.181	5.4
Cooling tower fans	15	0.181	2.7
Chilled water pumps	22	0.239	5.3
TOTAL	-	-	58.7

It is now possible to calculate the electrical demand, D_E, for the refrigeration plant and its auxiliaries.

$$D_E = 10^3 \times 58.7 \times 3.82/14000 \quad 16.0\ W/m^2$$

The figure 3.82 is the primary energy factor for electricity (2,3).

A similar calculation done for a pair of water-cooled vapour-absorption water chillers, for the same building and using steam at 0.84 bar gauge provided by a natural gas-fired boiler yields an electrical demand of 4.5 W/m² and a thermal demand of 35.0 W/m² making a total demand of 39.5 W/m² in primary energy terms to be compared with 16.0 W/m² for the vapour-compression system.

This is virtually as far as the Code has reached at the time of writing.

Future development

The immediate concern for future development is:

1 To repeat the examples mentioned for three other types of hypothetical building (see Part 2(a) of the Code) with different input parameters to establish a time for manual calculation and to test the methods further. This is necessary to determine the practicability of manual checking for any computer programs that may be produced.
2 To extend the examples mentioned to include the fans and other components of an induction system.
3 To establish a workable method for calculating mean monthly thermal demands.
4 To repeat the examples for other types of air conditioning system, particularly variable air volume.
5 To repeat (4) above, for at least two other types of building.
6 To review the results obtained and to consider the possibility of establishing a Building Envelope Number for air conditioned buildings.
7 To use the Building Envelope Number to establish thermal, electrical and total demand targets for air conditioned buildings.
8 Should (6) and (7) prove unsuccessful, to consider alternative methods of establishing targets, notably, the possibility of relating these to the performance indicators quoted in Part 4 of the Code.
9 To consider the development of software for a computerised approach.

There are still many problems to be solved but these appear, at least at the moment, to be responding to the approaches used. Optimistically, Part 2(c) of the Code could be published within two years.

Acknowledgements

The Code in all its parts could not have advanced to its present state without the considerable assistance provided by the Department of the Environment, both in manpower on the Working Party and by funding. The author and the Working Party are grateful to the Department. Thanks are also due to Ove Arup and Partners, Haden Young Ltd and North Thames Gas.

References

1 Peach J. *Energy Targets for Buildings, Experience of Energy Conservation in Buildings* (Ed. A.F.C. Sherratt), Construction Press pp. 10–24, 1981.

2 Building Energy Code: *Part 1, Guidelines Towards Energy-Conserving Design of Buildings and Services* — CIBS, 1977.

3 Building Energy Code: *Part 2, Calculation of Energy Demands and Targets for the Design of New Buildings and Services, Section (a) Heated and Naturally Ventilated Buildings*, CIBS 1981.

4 Building Energy Code: *Part 3, Guidance Towards Energy Conserving Operation of Buildings and Services*, CIBS 1979.

5 Building Energy Code: *Part 4, Calculation of Energy Consumption and Targets for Existing Buildings and Services*, CIBS 1982.

6 Traffic Noise and Overheating in Offices, *Building Research Establishment Digest* 162, February 1974.

4 Thermal Storage in Air Conditioning Systems

David C.J. Peters

Introduction

This chapter is intended to provide some information for the building systems engineer on the subject of Thermal Storage, or more accurately, Cool Storage, as applied to commercial air conditioning in the U.S.A.

The incidence of cool storage applications in the U.S.A. is on the increase. It is becoming established as a major new trend. There are thermal storage applications in other parts of the world, but nowhere are the number of applications increasing at the same rate as in the U.S.A., so this paper is confined to the design practices in that country. Also, the author's direct experience with the subject is in the U.S.A. only.

The sequence of the paper will be as follows:

1 The reasons for the trend toward thermal storage in the U.S. market are discussed.

2 A short overview of the principles of thermal storage, and the types of systems and buildings where cool storage is being applied in the U.S.

3 A more detailed explanation of a typical system design for one of the more promising approaches — ice storage. To include: system configuration, equipment operation, control concepts and equipment sizing.

4 A mathematical model which allows the designer to quickly determine the size relationship between storage and refrigeration equipment, for the minimum first cost. Also a discussion of other design considerations; energy usage, electrical demand and electric utility rate structure.

Market trends

Electric utility company situation

There are many electric utility companies throughout the U.S.A. that have their peak demand occurring during the summer months. While there are many items that make up the total demand in a building, it's often the air conditioning, or cooling equipment in both residential and commercial buildings that pushes that peak up to its highest point.

The power company has to have enough generating equipment available to meet this peak demand whenever it occurs, or their system will fail. As people and businesses continue their demand for more power, that means they have to construct new generating plants, which are currently at a very high cost. Although there are some shining examples to the contrary it is often a struggle to have them constructed on time, or within budget, due to the varied restrictions placed on power plant construction, because of environmental issues.

Also, in order for the electric utility companies to meet their demand at short notice, the peak usually has to be handled by smaller plants using gas turbines or similar equipment that can be put on the line quickly, and shut down again just as quickly, and these plants have a very high cost of operation, the fuel alone perhaps costing more than the average charge to the consumers per kwh.

This is illustrated by Figure 4.1. It's a typical daily demand curve for many summer peaking utilities. As the load builds during the day it can no longer be handled by the large generating plants that have the lowest operational cost, and the high cost, short term equipment has to be used to

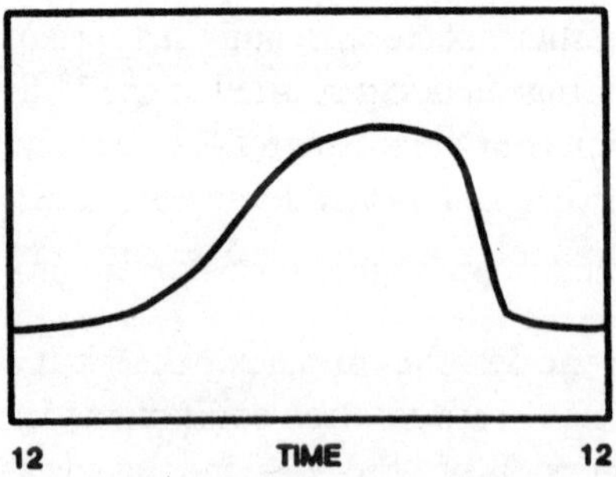

Figure 4.1 *Typical electric utility demand curve*

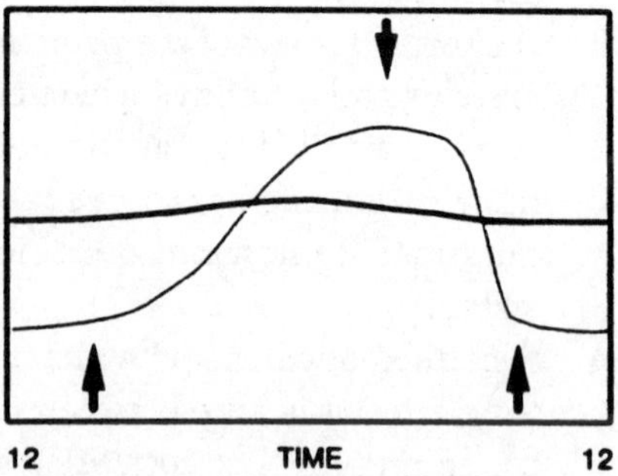

Figure 4.2 *Ideal electric utility demand curve*

handle this peak, which lasts through the evening meal and office clean up periods before cutting back to the basic night time load.

This represents a tough situation for the utilities, trying to manage their costs while continuing to provide reliable service.

Figure 4.2 represents a much more palatable situation for the utility, where the peak demand is reduced considerably, and instead the load is transferred to the night time. The net effect is an even curve, with the lower cost, base load plant being nearly fully loaded at all times, running at peak efficiency and the short term high cost plants never used. This curve is imaginary, it is unlikely that a demand curve such as this could be achieved, but it is shown in order to illustrate a desirable trend for the utility.

In order to encourage this trend, the utility company can offer a number of different incentives to their commercial customers, and certain utilities are offering one or more of the following:

1 Time of use rates, for both energy use and demand: — Normally a commercial customer pays a fairly large percentage of the electric utility bill for demand penalty, perhaps as much as 50% of the total bill, and the remainder is a direct charge for the energy used. A time of use rate usually means that there is a very much reduced or zero demand charge for *off peak* usage, and in some cases the energy use rate itself is reduced during off peak hours also.

2 Direct cash payment for *on peak* demand avoidance: — In some of the major cities, San Francisco, Dallas, and Boston for instance, commercial building owners and developers are being offered a direct payment (up to $300 per kw) for avoided demand. The utility company establishes a *norm* for the demand for a particular building, based on usage, size and type. If the building systems engineer can design systems that have a demand less than the *norm* then the utility company makes a payment to the owner during construction to help defray any additional capital costs of the demand reducing systems.

3 Participation in cost of design studies: — Because some of the more sophisticated new systems require more analysis and design time than conventional systems at present, some utility companies are willing to pay an owner a portion of the additional engineering costs involved.

4 Advertising and Promotion. Some utilities are conducting awareness campaigns to educate building owners and engineers to the benefits of specific systems such as thermal storage.

In summary, it is the Electrical Utility companies that are providing the powerful financial incentives that are driving the market in the direction of thermal storage.

Business situation

Unfortunately for the power companies the growth of thermal storage has been limited by the fact that most of the systems require greater capital cost at the time of installation, and even though short paybacks can be shown by studies, most U.S. business is short term oriented, so that thermal storage has not had the full acceptance that might be expected. In addition, like all new systems thermal storage is not yet fully accepted by the engineering design professionals.

Other influences

The most interesting new development in the market has been the use of a combination of ice storage and low temperature air distribution systems, that have been shown to actually cost less than conventional HVAC systems. If such systems gain wide acceptance then the growth of thermal storage can be expected to be at a much greater rate, because it will be almost independent of incentives provided by the power companies.

Thermal storage design principles

Effect of storage on electrical demand

The major electrical loads in an office building in approximate order of size are:

- Lighting
- HVAC
- Elevators
- Miscellaneous power
- Hot water

Of these the largest is lighting; even with today's design standards of less than 22 W/m^2 (2 watts/sq.ft.) it represents about 45 percent of the total demand. HVAC is the next major opportunity for demand reduction. The other loads (in an office building) are relatively minor, except perhaps miscellaneous power, when it includes today's personal computer loads, but they do not offer much opportunity for demand shifting.

The HVAC system electrical demand is made up of:

- Refrigeration compressors
- Air handlers
- Cooling towers
- Pumps

listed in descending order of magnitude.

The cooling load of the building causes the basic electrical demand and is made up of the following components, also in descending order of magnitude:

- Lights
- Ventilation
- Solar
- People
- Transmission

None of the above information is new to the system designer, on every project the cooling load is calculated for the peak design day, and the HVAC system is designed and equipment selected from this.

Load profile

For a conventional system, a peak *design day* load is sufficient, however to properly analyze a storage system requires more than this; a complete design day cooling load profile, which plots the load as it varies at each hour of the day. Such a load profile is fairly easy to obtain today using a computerized load program.

Figure 4.3 shows the building load profile for a 25,600 m^2 (275,000 sq. ft.) office building with a peak design load of 2461kW (700 tons). As the system is switched on, the load rises immediately to a value fairly close to the peak because of the *pull down* load requirements. As the pull down load diminishes, the load drops, then rises again to the peak in mid-afternoon, before dropping again. At the end of the day the system is switched off and the load reduces to zero.

What is shown is the system operating load profile, rather than a pure theoretical building load profile. A profile of the building for 24 hours

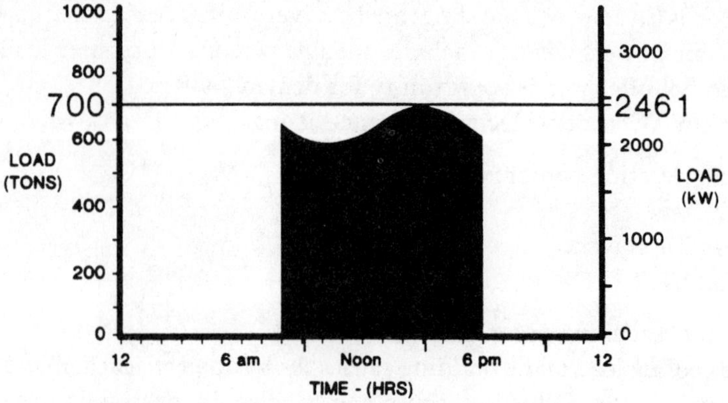

Figure 4.3 *Building load profile*

would show residual load in the building structure beyond the operating hours shown.

It is part of this residual load that the equipment has to deal with as *pull down* load. It is important when sizing a storage system that an assessment be made of this pull down load to ensure that the storage will be adequate, and is why the profile in Figure 4.3 indicates a substantial pull down load.

For any thermal storage system the total cooling energy requirement shown by the area under the *operating* load profile is considered when sizing the storage and refrigeration equipment. In this example the total requirement is 81,640 MJ (6450 ton hours).

System concepts

For maximum possible demand reduction, the system can be sized to store the total cooling energy requirement off peak, and keep the refrigeration equipment switched off during the day, using the stored cooling to offset the daytime cooling load. This is known as load shifting, or a *full* storage system. This type of system is simple to size, using the total cooling requirement; however both the refrigeration equipment and storage are of maximum size, so the cost of a *full* storage system is relatively high, making it impractical to use unless the utility rate structure is heavily in favor of off peak power usage. This is rarely the case, making *full* storage a relatively unpopular system. A full storage system load profile is shown in Figure 4.4.

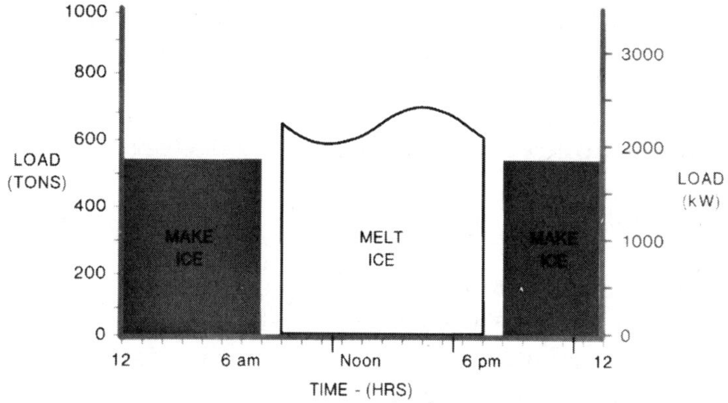

Figure 4.4 *Full storage load profile*

Another approach is to provide reduced storage, and run the refrigeration equipment during the day to handle the remaining portion of the cooling load. The same equipment is then used to store cooling energy at night. This is known as load levelling or *partial* storage.

A partial storage system requires some analysis to determine the correct size relationship between ice and refrigeration equipment, depending on the system configuration and the control strategy adopted.

A load profile for the 2461kW (700 ton) example office building showing the "partial" storage system concept is shown in Figure 4.5. By sharing the load approximately 50% between the melting ice and the chiller during the day, both the storage and chiller capacities are considerably reduced compared with full storage or with a conventional system. At the same

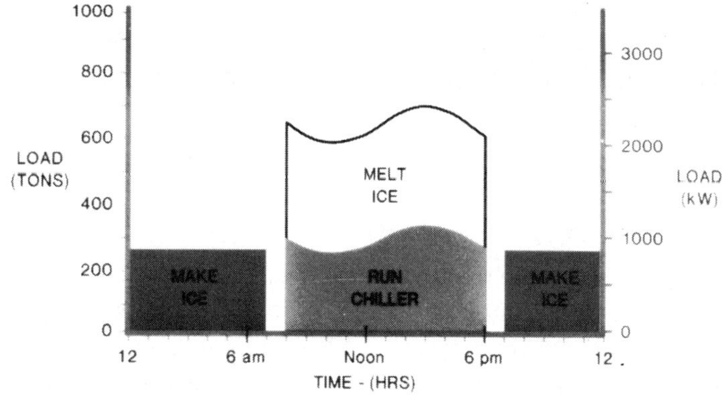

Figure 4.5 *Partial storage load profile*

time, the demand caused by the refrigeration is reduced by approximately 50% compared to a conventional "on peak" system.

Types of buildings and systems

Qualified storage buildings

By examining the load profile examples (Figures 4.3, 4.4, 4.5) it is readily evident that the most suitable buildings for thermal storage are those that have defined operating hours and a distinct peaking day time load, with zero night time load. Such buildings are:

- Offices
- Retail stores, shopping centers
- Schools, Universities

While it is possible to design a storage system for a building that has 24 hour cooling equipment operation, where cooling would have to be provided at the same time that storage is being generated, the system becomes more complex, and there is not usually sufficient financial advantage to make such a design worthwhile, so such buildings are generally less suitable for thermal storage. They include:

- Hospitals
- Hotels
- Apartment Buildings

Many of the more visible storage projects are on medium or large buildings, but a number are being designed and installed for smaller projects, down to approximately 200 m^2 (22,000 sq.ft.) total floor space which equates to a cooling load of approximately 175kW (50 tons). So thermal storage is being applied over a wide range of commercial building sizes.

Storage methods

At present, thermal storage in air conditioning is only being installed on applied type central systems, not on unitary or packaged equipment projects, although one U.S. manufacturer has just announced a rooftop package for ice storage of only 18kW (5 tons) nominal capacity, and several Japanese manufacturers have such small units in their domestic market.

The actual medium used for storage varies, and includes the following:

- Chilled water
- Ice
- Eutectic Salts
- Slurries
- Clathrates (Freon hydrates)

The last three mentioned are generally in the experimental stage and have been used on pilot projects only for comfort cooling, most applications have used chilled water or ice.

Chilled Water. Chilled water storage projects have been done all over the world, a principal centre of activity has been Canada, where a number of projects have been in operation for some years. A major disadvantage is that large tanks are required to store sufficient chilled water. The 2461kW (700 ton) office building example mentioned earlier would need approximately 2,300 m^3 (500,000 Imp. gallons) of chilled water for a full storage system, however an advantage of chilled water storage is that the refrigeration equipment can be standard water chillers.

A major design challenge with chilled water is to ensure that the water charging the tanks does not mix with the discharged water, or the storage system becomes inefficient, and this has caused many different design approaches, distribution headers, baffles, compartmented tanks that are connected in series, and one method uses a moving flexible membrane in the tank to keep the two water streams separated.

Some economical installations have been done in Canada with both cooling and heat storage, in combination. Using multi-celled tanks interchangeably as the load varies from summer to winter, and reasonable paybacks have been achieved, but operational problems are much greater than for conventional systems.

Ice. The use of ice as a storage medium has one obvious major advantage over chilled water; the space required is approximately 10 times less due to the use of latent heat instead of only sensible heat.

The major disadvantage is that the refrigeration equipment must produce lower temperatures than conventional chilled water systems, and therefore requires greater power input to the compressor. This is offset somewhat by the fact that ice is generated at night only, when the outdoor temperature is lower.

Low air temperature systems

As mentioned earlier, it has been found advantageous to use the available lower temperatures of an ice storage system to produce a lower supply air temperature to the building: 5C (41F) instead of the more conventional 15C (55F).

The lower supply air temperature results in almost 50 percent less airflow in the central air handling system, with significant size and cost reduction of the air handling units and the ductwork distribution system.

In addition the fluid temperatures used have a wider temperature difference between supply and return to the cooling coils, on the order of 11K (20F) instead of the more conventional 6K (11F), so a similar size and cost reduction is also possible in the fluid piping distribution and pumps.

The low supply air temperature of 5C (41F) is not supplied directly to the air conditioned space, but blended with return air from the space in the ceiling plenum by means of a fan powered mixing box terminal. By this method air distribution and circulation rates are the same as a conventional system.

The combination of a low supply air temperature, high fluid temperature difference and *partial* ice storage system has been found (by studies, and in a few installations) to have a lower total installed cost than a conventional central chilled water/VAV system.

In addition the combination of *partial* storage system, (with smaller refrigeration equipment) and reduced fan and pump requirements results in a significant daytime eletrical demand reduction overall for the HVAC system in a building.

While this type of system has not yet gained universal acceptance, it has good potential, therefore the remainder of this chapter uses such a system to illustrate the principles of system design, control and equipment sizing for thermal storage.

System design

For simplicity of presentation the system examples shown here use brine (usually 25% ethylene glycol/water) as the fluid to generate ice, and to circulate cooling to the air system, therefore the refrigeration equipment is referred to as a chiller. Many of the principles shown however apply equally well to other types of systems, such as direct refrigeration for ice generation and to chilled water storage systems. All the examples are *partial* storage.

System configuration

To operate a partial storage system effectively, it is necessary to arrange the piping between the ice storage and chiller in such a way that they can both handle their assigned portion of the load.

A simplified partial storage arrangement is shown schematically in Figure. 4.6. Block 1 represents the ice storage tanks, in series with the chiller represented by Block 2. Block 3 represents the building load, in the form of the air handling unit coils. In this case brine piping connects the equipment, with flow direction shown by the arrows. During the night time, flow to the building load is shut off, and the chiller operates at low temperature to form ice in the tanks (1).

During the day time, the building load is provided with cooling from both the melting ice bank and the refrigeration equipment, in effect the ice bank is now acting as a chiller, in series with the brine chiller equipment, in this way the full cooling load is provided. This is described as a *chiller downstream* arrangement.

By reversing the positions of ice bank and chiller, a *chiller upstream* arrangement is obtained. Both *upstream* and *downstream* provide the same cooling performance as far as the building load is concerned. The individual performance of the ice bank and chiller is different in each case however, and each arrangement has important advantages and disadvantages which affect the cost and energy use of the overall system. It is also possible to configure the system so that chiller and ice bank operate in parallel. A parallel arrangement also has some advantages when used with conventional higher daytime system temperatures.

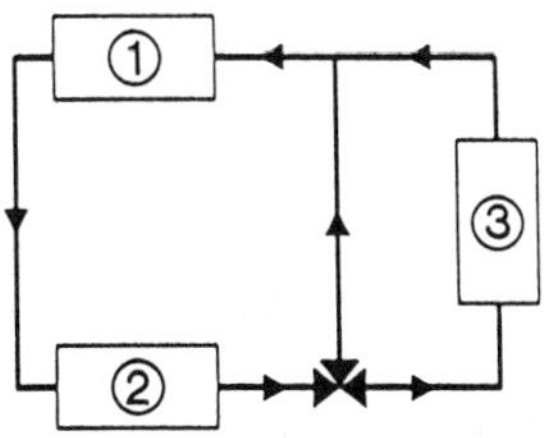

Figure 4.6 *Partial storage system schematic*

Equipment operation

To understand the advantages of each system arrangement it is necessary to study the performance of the major system components **Chiller** and **Ice Bank** individually.

To illustrate the performance of each component in different system configurations, typical brine temperatures have been added to the simplified schematic diagram in Figure 4.7. During ice making mode (night time) the chiller provides – 4C (25F) supply temp with – 1C (30F) return temp. (Both supply and return temperatures must be below 0C (32F) to freeze the water in the ice bank). During the ice melting mode (day time) the building load, or air cooling coils are supplied with brine at 2C (36F) and the return temperature from the building is 13C (56F) (an 11K (20F) temperature difference). If Blocks 1 and 2 are each providing approximately 50% of the day time load then the temperature differences across each must be 5.5K (10F), the intermediate temperature is therefore 7.5C (46F) between Block 1 and 2.

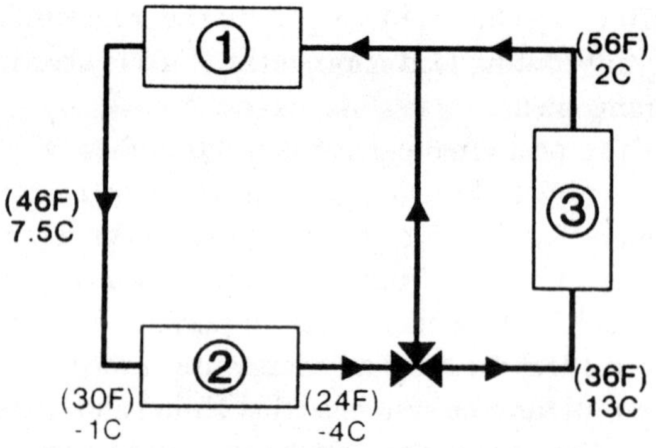

Figure 4.7 *System schematic with typical operating temperatures*

Chiller

The chiller is first selected for ice making (night) duty, – 1C (30F) entering, – 4C (24F) leaving the evaporator) because this is the duty at which the compressor has to handle the highest *lift*, or temperature difference between the evaporator and condensor. At this duty the chiller has its lowest capacity.

If the chiller is operated in the *day* mode in the downstream configuration (i.e. in Block 2), the evaporator operates with 7.5C (46F) entering and

2C (36F) leaving brine. The same chiller will now perform at higher capacity than in *night* mode.

When the same chiller is operated in the *upstream* position, (Block 1), the evaporator entering temperature is 13C (56F), leaving at 7.5C (46F), giving still greater capacity than the *downstream* position.

The specific power input per ton of capacity is greatest for night-time ice making duty and least for the daytime upstream position, although the actual total power input at maximum capacity is actually higher in the *day* mode than the *night* mode.

In general then, an *upstream* chiller requires the least energy input per ton of capacity, and also, the energy required per ton of ice making capacity is greater than the energy required per ton of *day* duty.

Ice Bank

When the ice bank is in the *downstream* position (Block 2) the leaving temperature of 2C (36F) is close to the melting temperature of 0C (32F) therefore there is less capacity available in the tank, and the time in hours to discharge the tank is less. The same tank in the upstream position has a greater capacity available, because the leaving temperature is at a greater difference from the 0C (32F) ice melting temperature than a downstream tank. The time available to completely discharge the tank is proportionately longer. It is also easier to control the ice storage leaving temp throughout the cycle when the temperature difference between melting temp and leaving temp is greater.

The case of parallel operation should also be considered. A schematic diagram, with typical operating temperatures is shown in Figure 4.8. Note

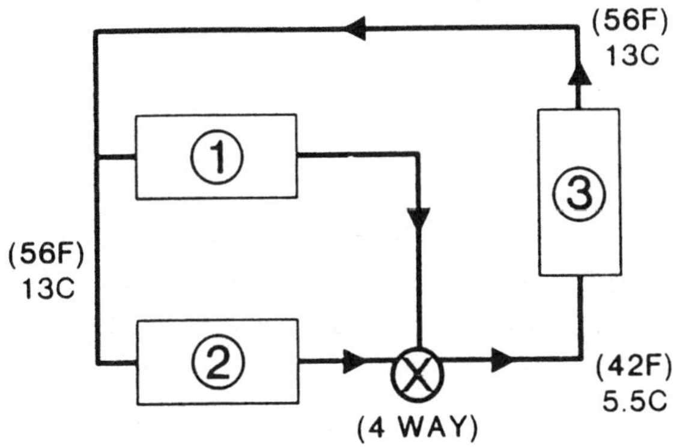

Figure 4.8 *System schematic (parallel)*

that this arrangement should be used at conventional system temperatures, so that the chiller and ice bank can both be operated at temperatures similar to the *upstream* mode, with the resulting advantages for both ice bank and chiller operation.

Control concepts

As well as system configuration, the designer needs to consider the method which will be used to control the chiller and ice storage. The control concept chosen also affects system performance and equipment sizing.

There are three basic methods used to control a partial ice storage system: **Chiller priority**, **Ice priority**, and **Constant proportion**. The three methods can best be considered by studying the different load profiles, as in Figures 4.9, 4.10 and 4.11.

Chiller priority

This term refers to the fact that during *day* mode the chiller handles the base load of the building, at a fairly constant rate, while the ice bank is used to handle any variations in the load. This applies also for lower loads other than the design day. The chiller is therefore used for the majority of operating hours to handle the load, and the resulting electric demand is constant throughout the year. The ice bank is used only for *topping up*; therefore the chiller night time use to regenerate the ice bank is minimised also, on other than design days.

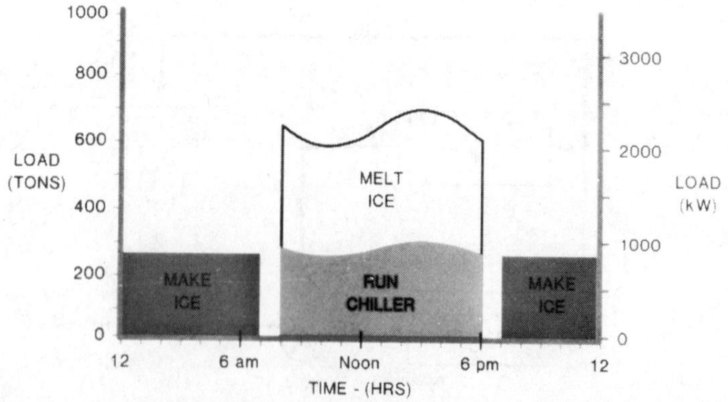

Figure 4.9 *Partial storage profile I — Chiller priority*

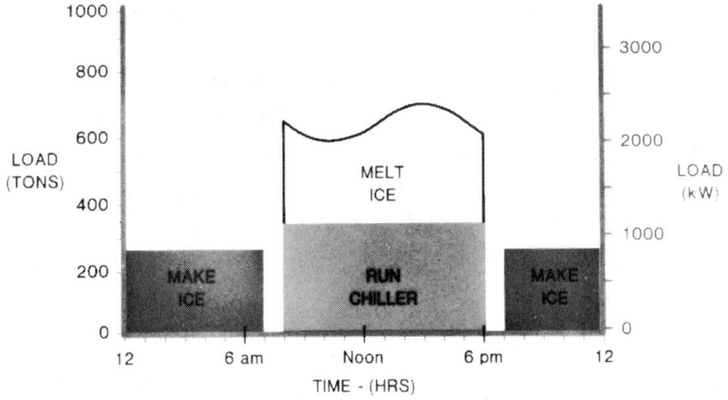

Figure 4.10 *Partial storage profile II — Ice priority*

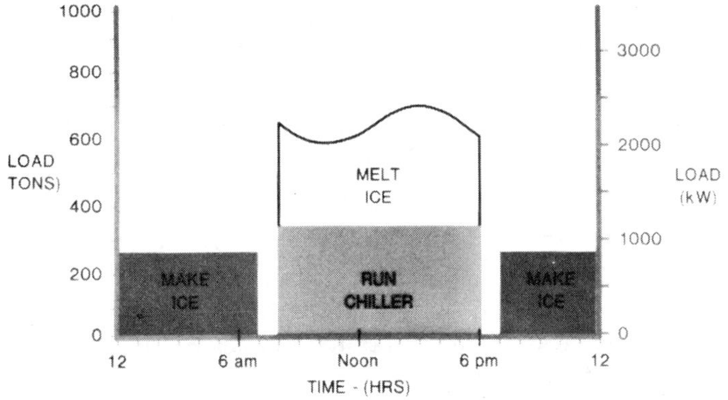

Figure 4.11 *Partial storage profile III — Constant proportion*

This particular control strategy is the best one of use when the night time energy use rate is relatively high, because the more energy intensive ice making mode is minimized. It is also best to use when there is a *ratchet* demand clause in effect, since under these conditions there is no gain from reducing the day time chiller base load and therefore reducing demand below a minimum. (*Ratchet* demand refers to the type of rate where the maximum demand attained during the cooling season sets the demand charge for each succeeding month).

This method is the simplest to install and operate.

Ice Priority

In this case, the ice bank is used as the base load, and the chiller for *topping up*. The use of the ice bank is therefore maximized, and the chiller use minimized during the *day* mode.

This control strategy is best used when there is a time of use energy rate in effect, making the system less expensive to operate during the ice regeneration mode at night. It also minimizes the daytime demand, so is most effective when the rate structure has a heavy demand penalty, without a *ratchet* clause.

Ice priority is relatively simple to install and operate but not quite as simple as chiller priority.

Constant proportion

This method is a compromise between the chiller and ice priority methods, maintaining a constant proportion between the portion of the loads handled by the ice bank and the chiller, at all load conditions. This method is most adaptable to any rate structure, and can be used to optimize the operation when a complex rate structure is in effect.

This is the most complex of the three control schemes in installation and operation.

Equipment sizing

Having discussed the various options and decisions that the system designer is faced with, it is necessary to get to the main objective of this chapter, and explain how the equipment can best be selected for various combinations of these options and still minimize the installed cost of the equipment.

As was discussed earlier in the Equipment Operation section of this chapter, the capacity available from a given chiller is less during the *night* than in the *day* mode. A major assumption in developing the sizing formulae given in this chapter, is that is order to minimize the total cost of the system, it is necessary to fully utilize the maximum available chiller capacity during both *night* and *day* modes, so that there is no excess capacity available during either mode. If the capacity available from the chiller were equal in both modes, then the task would be simple. Also in a partial storage system, equal capacity in both modes would result in the greatest demand reduction, so the ideal night/day capacity factor F_{ch} for demand reduction is 1.0. By the definition, a given chiller has an individual capacity factor that is always less than 1.0. The factor varies with the type of chiller used, individual machine selected, and the particular night/day operating temperatures dictated by the system configuration.

Definition of variables

To present the sizing formulas, it is first necessary to define all the other variables, in addition to F_{ch}. Figure 4.12 is a load profile for the simplest example, chiller priority operation, and together with Table 4.1 illustrates each variable and the units of measurement used for calculation. Figure 4.12 is shown with the hours of operation shifted for clarity, beginning with night chiller operation, rather than the conventional midnight to midnight load profile. From a practical standpoint, it is important to note that the capacity values obtained by the calculation methods shown here are the minimum theoretical values, with no allowance for safety factors.

Table 4.1 *Variables*

L	= TOTAL COOLING - DESIGN DAY	kJ	(TON HRS)
L_I	= ICE STORAGE PORTION OF TOTAL COOLING	kJ	(TON HRS)
L_C	= CHILLER PORTION OF TOTAL COOLING	kJ	(TON HRS)
Q_P	= PEAK TOTAL BUILDING COOLING LOAD	kW	(TONS)
Q_A	= DESIGN DAY AVERAGE BUILDING COOLING LOAD	kW	(TONS)
Q_I	= ICE STORAGE MAXIMUM INSTANTANEOUS LOAD	kW	(TONS)
Q_D	= CHILLER CAPACITY (DAY)	kW	(TONS)
Q_N	= CHILLER CAPACITY (NIGHT)	kW	(TONS)
D	= DAY OPERATING TIME	s	(HRS)
N	= NIGHT OPERATING TIME	s	(HRS)

$$L = L_I + L_C$$

$$Q_P = Q_I + Q_D$$

$$F_B = \frac{Q_A}{Q_P}, \quad F_{CH} = \frac{Q_N}{Q_D}, \quad F_S = \frac{L_I}{L}$$

Sizing formulas

Chiller priority

In the case of a chiller priority system, because the chiller is operating at constant load during the night time, and a different, but constant, load

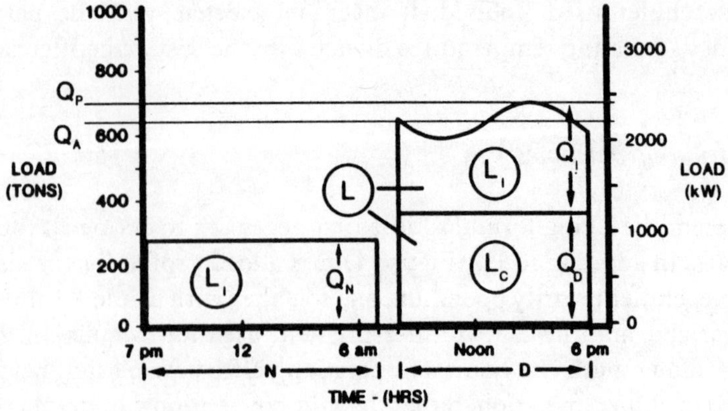

Figure 4.12 *Load profile—variables*

during the day time the sizing formula is straightforward to develop, as follows:

$$L = L_C + L_I \qquad \textbf{(equation 1)}$$

$$L_C = D \times Q_D, \; L_I = Q_N \times N$$

Substitution:

$$L = (D \times Q_D) + (Q_N \times N) \qquad \textbf{(equation 2)}$$

By definition $F_{CH} = \dfrac{Q_N}{Q_D}$

Therefore: $Q_N = F_{CH} \times Q_D$

Substituting for Q_N in equation 2

$$L = (D \times Q_D) + (F_{CH} \times Q_D \times N)$$

$$L = Q_D \,[D + (F_{CH} \times N)\,]$$

Solving for Q_D

$$Q_D = \frac{L}{D + (F_{CH} \times N)} \qquad \textbf{(equation 3)}$$

Ice priority

In an ice priority system, the chiller day load is not constant, but varies in proportion to variations in building load, it is therefore necessary to introduce a term which accounts for this variation (F_B)

By definition:

$Q_P = Q_D + Q_I$

therefore $Q_D = Q_P - Q_I$ **(equation 4)**

also by definition:

$$F_B = \frac{Q_A}{Q_P}\text{, therefore } Q_P \frac{Q_A}{F_B} \qquad \textbf{(equation 5)}$$

$$\text{Also } L = Q_A \times D\text{, therefore } Q_A = \frac{L}{D}$$

substituting for Q_A in equation 5

$$Q_P = \frac{L}{F_B \times D}$$

substituting for Q_P in equation 4

$$Q_D = \frac{L}{F_B \times D} - Q_I \qquad \textbf{(equation 6)}$$

$$L_I = Q_I \times D = Q_N \times N \qquad \textbf{(equation 7)}$$

$$\text{by definition } F_{CH} = \frac{Q_N}{Q_D}$$

therefore $Q_N = F_{CH} \times Q_D$

substituting for Q_N in equation 7

$Q_I \times D = F_{CH} \times Q_D \times N$

$$\text{therefore } Q_I = \frac{F_{CH} \times Q_D \times N}{D}$$

substituting for Q_I in equation 6

$$Q_D = \frac{L}{F_B \times D} - \frac{F_{CH} \times Q_D \times N}{D}$$

$$Q_D + \frac{F_{CH} \times Q_D \times N}{D} = \frac{L}{F_B \times D}$$

$$Q_D \frac{(1 + F_{CH} \times N)}{D} = \frac{L}{F_B \times D}$$

$$Q_D = \frac{L}{(F_B \times D)\left[1 + \frac{F_{CH} \times N}{D}\right]}$$

$$Q_D = \frac{L}{F_B [D + (F_{CH} \times N)]} \qquad \textbf{(equation 8)}$$

Constant proportion

For constant proportion control, by definition the proportion of load handled by the ice and that handled by the chiller have a fixed relationship. This can be stated as follows:

$$F_S = \frac{L_I}{L} = \frac{Q_I}{Q_P}, \text{ or } \frac{L_C}{L} = \frac{Q_P}{Q_P}$$

$$\text{therefore } L_C = L \times \frac{Q_D}{Q_P} \qquad \textbf{(equation 9)}$$

From equation 7 earlier in the text

$$L_I = Q_N \times N$$

$$\text{by definition } F_{CH} = \frac{Q_N}{Q_D},\ Q_N = F_{CH} \times Q_D$$

substituting in equation 7

$$L_I = F_{CH} \times Q_D \times N \qquad \textbf{(equation 10)}$$

restating equation 1

$$L = L_C + L_I$$

substituting equation 9 and equation 10 into equation 1

$$L = \frac{(L \times Q_D)}{Q_P} + (F_{CH} \times Q_D N)$$

$$= Q_D \left[\frac{L}{Q_P} + (F_{CH} \times N) \right]$$

solving for Q_D

$$Q_D = \frac{L}{\frac{L}{Q_P} + (F_{CH} \times N)}$$

since $L = Q_A \times$, substituting for L

$$Q_D = \frac{L}{\frac{Q_A \times D}{Q_P} + (F_{CH} \times N)}$$

by definition $F_B = \frac{Q_A}{Q_P}$

therefore $Q_D = \frac{L}{(F_B \times D) + (F_{CH} \times N)}$ **(equation 11)**

Comparisons

By comparing these equations for Q_D for each of the three control methods it is possible to determine the relationships for equipment sizing.

Restating equations 3, 8 and 11

$Q_D(CP) = \frac{L}{D + (F_{CH} \times N)}$ (CP denotes Chiller Priority)

$Q_D(IP) = \frac{L}{F_B [D + (F_{CH} \times N)]}$ (IP denotes Ice Priority)

$Q_D(K) = \frac{L}{(F_B \times D) + (F_{CH} \times N)}$ (K denotes Constant Priority)

Comparison A — Compare $Q_D(CP)$ to $Q_D(IP)$

$$\frac{Q_D(CP)}{Q_D(IP)} = F_B$$

So it can be stated that the chiller size for a chiller priority system is

$1 - (F_B \times 100)$ % smaller than the chiller for ice priority.

Comparison B — Compare $Q_D(K)$ to $Q_D(IP)$

$$\frac{Q_D(K)}{Q_D(IP)} = \frac{F_B [D + (F_{CH} \times N)]}{(F_B \times D) + (F_{CH} \times N)}$$

$$= \frac{(F_B \times D) + (F_B \times F_{CH} \times N)}{(F_B \times D) + (F_{CH} \times N)}$$

$$= \frac{F_B \times F_{CH} \times N}{F_{CH} \times N}$$

Since F_B is less than 1.0 $\frac{Q_D(K)}{Q_D(IP)}$ is also less than 1.0.

The statement can therefore be made that the chiller for Constant Proportion control is smaller than that for ice priority.

Comparison C — Compare $Q_D(K)$ to $Q_D(CP)$

$$\frac{Q_D(K)}{Q_D(CP)} = \frac{D + (F_{CH} \times N)}{(F_B \times D) + (F_{CH} \times N)}$$

Since F_B is less than 1.0, we conclude

$\frac{Q_D(K)}{Q_D(CP)}$ is greater than 1.0 and therefore, it can be stated that the chiller for Constant Proportion control is larger than that for chiller priority.

Size Relationship

Since $L_I = Q_D \times F_{CH} \times N$, and $F_{CH} \times N$ is the same for all systems, storage tank size and chiller size follow the same relationship for the different control methods. The size relationship may be summarized as follows:

Both chiller size and size and storage size are smallest for chiller priority, and largest for ice priority. For constant proportion the size of chiller and storage is greater than chiller priority, but less than ice priority.

Example calculations

To use these formulas in practice, it is necessary to assume a value for F_{CH} to make an initial sizing calculation. After selecting a chiller, check that it's actual individual value for F_{CH} is less than the one initially selected for calculation.

Table 4.2 gives a range of typical minimum F_{CH} values for chiller equipment, to assist in the calculation process. The values given are for the

Table 4.2 *F_{CH} Values*

	RECIPROCATING CHILLER	CENTRIFUGAL CHILLER
CHILLER UPSTREAM	0.65	0.60
CHILLER DOWNSTREAM	0.80	0.70
PARALLEL	0.65	0.60

temperature conditions given previously in this chapter (these will be found close to those used in practice) and for various chiller types and system configurations.

Note that the F_{CH} values listed for Parallel System configuration are similar to those for Chiller Upstream. This is because the operating temperatures are similar.

In addition to the comparisons between the various formulas, made earlier, a series of calculations was performed to size both chiller and ice bank for the example office building described earlier, using known values as follows:

Q_P = 2461kW (700 tons)

L = 81,640 MJ (6450 ton hrs.)

N = 12 hrs.

D = 10 hrs.

and various values of F_{CH}.

These calculations were performed using equations 3, 8 and 11 from this chapter and other simple basic relationships to obtain the day and night time chiller capacity, storage size and percent storage.

(Note: In order to obtain chiller capacities in kW, using the sizing formulas, the correct units for storage (L) are kJ, and the time must be converted from hours to seconds).

The results of these calculations are given in Table 4.3. By inspection of these results it is possible to further verify the statements

Table 4.3 *Calculation results*

CONTROL METHOD	F_{CH}	Q_D kW	Q_D (TONS)	Q_N kW	Q_N (TONS)	L_I MJ	L_I (TON HRS)	F_S X 100 (%)
CHILLER PRIORITY	0.6	1319	375	791	225	34171	2700	42
	0.7	1233	350	863	246	37282	2952	46
	0.8	1157	329	926	263	40003	3156	49
	0.9	1090	310	981	279	42379	3346	52
ICE PRIORITY	0.6	1432	408	859	244	37109	2928	46
	0.7	1338	381	937	267	40478	3204	50
	0.8	1256	358	1005	286	43416	3432	53
	0.9	1184	337	1065	303	46008	3636	56
CONSTANT PROPORTION	0.6	1382	393	829	236	35183	2832	44
	0.7	1287	366	901	256	38923	3072	48
	0.8	1205	343	964	274	41645	3288	51
	0.9	1133	323	1019	290	44021	3480	54

made in the sizing formulas given earlier, and to draw some further conclusions.

Table 4.4 *Cost ranking chart*

CONTROL METHOD COST PARAMETER	CHILLER PRIORITY	ICE PRIORITY	CONSTANT PROPORTION
LOWEST CHILLER COST	1	3	2
LOWEST STORAGE COST	1	3	2
LOWEST DEMAND CHARGES (RATCHET)	1	3	2
LOWEST DEMAND CHARGES (NON RATCHET)	3	1	2
LOWEST ENERGY COST (FLAT RATE)	1	3	2
LOWEST ENERGY COST (TIME OF USE)	3	1	2

Conclusions

In addition to the results given in Table 4.3, Table 4.4 shows a rough cost ranking that summarizes the statements made throughout this chapter. In Table 4.4a (1) denotes the best and (3) denotes the worst.

Table 4.4 lists, in groups, the three major parameters that a designer of a storage system is concerned with: Equipment First Cost, Demand and Energy Cost. The first two lines of Table 4.4 have been documented at some length in this chapter, the Demand and Energy cost rankings are generalizations based on studying several actual project designs, and are intended to give a guide to the design engineer in making his system selection.

To verify the summary statements made at the end of the Sizing Formulas section, look at a particular value of F_{CH} in Table 4.3. – 0.8, and the resulting values for Q_N (the capacity requirement used to size the chiller), and L_I. The ice priority value for Q_N, – 1005kW (286 tons) is 9% larger than that for chiller priority, – 926kW (263 tons) as is the storage requirement. The cost will therefore be proportionately greater for ice priority.

The effect of system configuration can also be significant. Looking at Table 4.2 F_{CH} values for a reciprocating chiller, if we use the chiller in a downstream configuration we must use a capacity factor of 0.8 or greater, in the upstream position we could use an F_{CH} value as low as 0.65 if desired.

Looking again at Table 4.3, chiller priority, the resulting values for Q_N would be: (for F_{CH} = 0.8) 926kW (263 tons) and (for F_{CH} = 0.65) 827kW (235 tons) (by interpolation). Therefore the potential difference in size is a 12% greater requirement for the downstream configuration.

It would appear therefore that chiller upstream is the desired configuration, however in actual practice, it is unlikely that a designer would use an F_{CH} value as low as 0.65 because the resulting demand reduction would be too small.

Another way to consider the size comparison is to look at a particular size of storage and chiller — (see Table 4.4 values of 49% (chiller priority) and 50% (ice priority) which are approximately the same). The chiller priority system uses less of the chiller daytime capacity, so will result in lower daytime demand, and will have a greater daytime safety factor built in.

The other considerations the design engineer must weigh carefully are the effect of system configuration, control concept and utility rate structure on the demand and energy use charges.

For example, when there is a non-ratchet demand, coupled with a time of use rate in effect, the design engineer will probably find that the extra first cost of the ice priority system is justified. Each system requires careful analysis of these factors on a project by project basis.

References

1 Equipment Sizing Concepts for Ice Storage Systems D. Peters, W. Chadwick, J. Esformes. *EPRI/CEA International Load Management Conference Proceedings* (Chicago USA, Dec. 9–11, 1985).

5 Good Practice in the Design and Construction of Air Conditioning Systems

Colin Swain

Introduction

A more accurate title for this chapter might have referred to Bad Practice rather than Good Practice in the Design and Construction of Air Conditioning Systems, as the purpose is to assist those concerned with new work by analysing and drawing conclusions from jobs that have gone wrong so that repetition of the same mistakes can be avoided.

Mostly the shortcomings are obvious with hindsight. Errors occur which no competent engineer would make in the cool light of day, and much of the advice given here appears trite and self-evident. However many of the examples quoted are of essentially good designs prepared by competent air conditioning engineers and installed by reputable contractors which went wrong because someone overlooked an elementary point. Sometimes this was because the person responsible took his eye off the ball or was too busy to see what was being done by his less experienced staff. In other cases, fundamentals were obscured by a mass of detail or the development time in a large project. On occasions the mistake was in not taking account of developing technology and other people's experience. The intention is to draw some lessons from all these cases to point the way to recognition of likely problem areas before troubles arise.

The case histories are of jobs with which I have been personally acquainted, so the range of failures is limited to those within my experience. They span a period of time and in some cases the technology has since changed although the lessons to be learned remain the same. The term air conditioning is interpreted somewhat broadly. For obvious reasons the identity of the projects and the firms involved is concealed.

Good practice

Good practice in the design and construction of an air conditioning system can be regarded as the procurement of a system that is well matched to its application in terms of — ability to maintain appropriate environmental conditions; harmonising with the building it serves; compatibility with the needs and facilities of the users; effect on external environment; capital and operating costs; space taken up; and serviceability, reliability and operating life. A compromise always has to be made between these often conflicting factors, and good practice is selecting the optimum design which provides the best balance in the particular circumstances.

The processes entailed in designing and constructing an air conditioning system which are considered in turn are —

1. Establishing requirements.
2. Preparing a conceptional scheme.
3. Ensuring suitability/compatibility with the building.
4. Detailed design and equipment selection.
5. Installation.
6. Commissioning, testing and handing over.

Establishing requirements

It is axiomatic that no design can start until the designer knows what task the air conditioning plant is to perform. The needs of the building owner or user vary with individual circumstances, so the first essential is to find out what the requirements are. The initial source of information must be the client, user or developer. Usually clients have a fair but incomplete idea of what they want, and it falls to the designer to establish not only what the client thinks he requires but what he actually needs. This entails skill and experience by the designer to ascertain exactly how the client intends to use the building and its air conditioning plant. It is not just a matter of asking the building owner or user — he probably doesn't know, particularly if he is not asked the right questions. Determining the operating parameters must be a two-way process in which the designer takes a leading role.

An example of the designer failing to fully understand the user's requirements occurred with the construction of a new processing room in a food factory. About 90 people were to work in the room, in which perishable food was taken in bulk from cold store and sorted, weighed and packaged before return to cold store. The processing room had to be held at a temperature of 10°C. The consultants sent out a detailed questionnaire to ascertain the user's requirements, from which they concluded that the two main operating regimes were steady state operation during production and starting up after prolonged shut-down. The system was designed and installed on this basis, with a two-speed air handling plant. However the plant consistently failed to hold space temperature and stock was being spoiled. The reason proved to be that to maintain the processing room scrupulously clean it was washed down three times a day with very hot water; and although the designers had known this from the answers to their questionnaire and had allowed for supplying the hot water demand they had not associated it as a load on the air conditioning plant. In fact moisture removal during wash-down was the biggest cooling load, and the inability to deal with it was the cause of the poor plant performance.

In another case, the requirements were established but misinterpreted by the designer. A five storey office building was being refurbished to accommodate the main computer of a London finance house. The central air conditioning plant had ducted supply to all floors with ceiling distribution from local VAV units. The computer manufacturer told the air conditioning designer the total installed cooling load and stated the anticipated diversity to be 60%, which was used as the design basis for the air conditioning plant. When the computer was installed, the air conditioning system was unable to hold the correct temperature. Investigation showed there had been a design error. Instead of applying the diversity factor only to the total load of the central plant with a progressive increase of simultaneous use to 100% at individual terminal units, the designer had applied the 60% factor throughout the plant so that the entire air distribution system was undersized. The only feasible remedy was to install freestanding air conditioning units at a cost of £250,000 and loss of useful floor space.

When agreeing requirements for the air conditioning, understanding must also be reached as to cost limitations. This is where the designer must be alert, particularly if asked to reduce his cost estimates or to work within a cost norm established from other not necessarily comparable schemes. The experienced designer will cut service rather than quality as a building fully air conditioned to a poor standard is never as satisfactory as a building partly air conditioned to a good standard. It is essential to state the consequences of proposed cost cuts and what the operating limitations will be.

Poor quality air conditioning schemes put in to an unrealistic price by a designer endeavouring to keep his client happy are too numerous.

Conceptional design

The next step is to implement the agreed requirements in terms of a conceptional design. This should involve the entire design team as the air conditioning both influences the building form and is governed by it, and even with existing buildings where the shape and essential characteristics are predetermined the air conditioning designer must take account of disciplines other than his own. Some of the points to be taken into account in the conceptional design of the air conditioning system are — building location; construction; shape; height; orientation; solar shading; open plan, cellular, or flexible layout; core arrangement; suspended ceiling; false floors; type and extent of glazing; and possible locations for central plant.

Preparing a conceptional design that takes account of these factors and also matches the requirements and purse of the client demands a great deal of skill and experience of the designer, and it is more an art than a textbook science.

A single-storey office was built over an existing computer building. Air conditioning was by unitary heat pump system, with floor units around the perimeter on full recirculation and two fresh air ceiling units discharging though diffusers. The installation was not satisfactory, people near windows complaining of cold draughts and those in the centre being warm and stuffy except for people under diffusers who suffered from cold air dumping. In my opinion the design was badly conceived in several respects. A 16m deep building has a significant core area and the cooling load cannot satisfactorily be handled entirely from the perimeter Figure 5.1(a). The perimeter units endeavouring to project cool air to the core promote poor air distribution, with stagnant pockets at the centre and vigorous air movement by the windows. Control problems with the ceiling units were inevitable, being on full fresh air and with wrongly positioned room thermostats. The ceiling unit fans are rated for limited external resistance and were stretched to deal with fresh air filter, preheater and ductwork on the suction side together with silencer, ducts and diffusers on the discharge side. A more experienced designer might have prepared a design like that in Figure 5.1(b), with a large core area served by a plant of ample capacity handling a mixture of fresh and recirculated air leaving the perimeter units

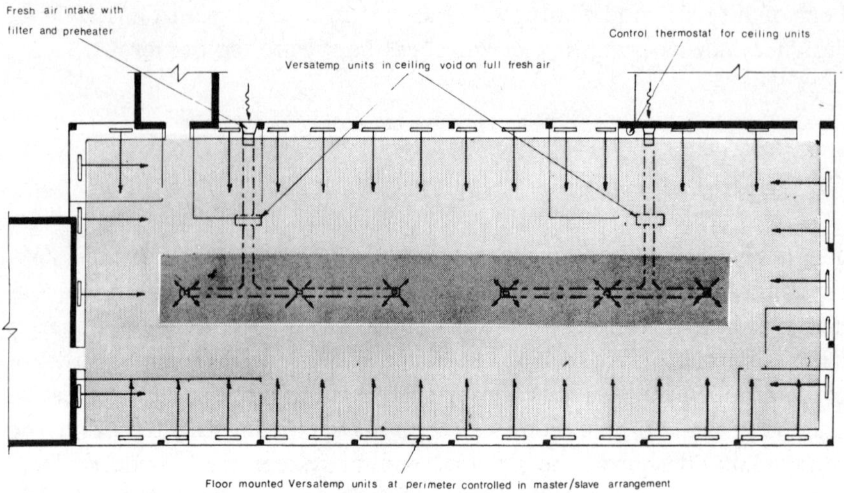

(a) As designed and installed

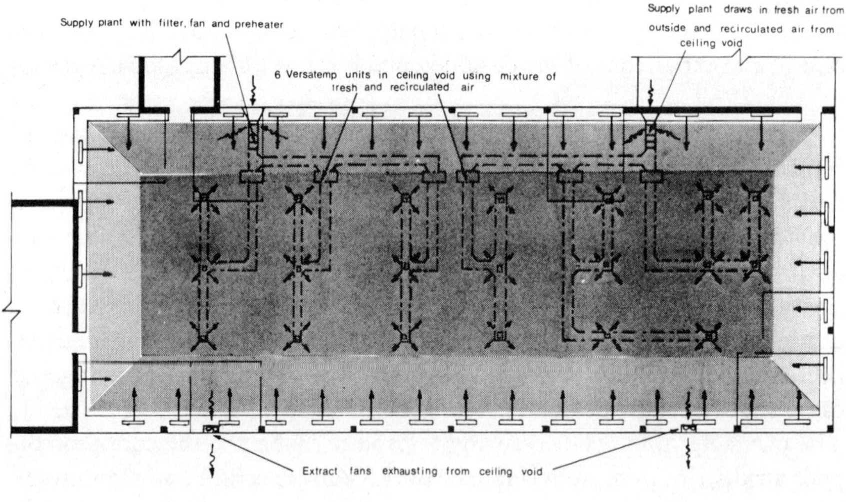

(b) As it should have been designed

Figure 5.1 *Air conditioning system in single-storey office*

to handle the cooling load of the walls and windows on their aspect together with just a proportion of general heat gain arising in their immediate vicinity.

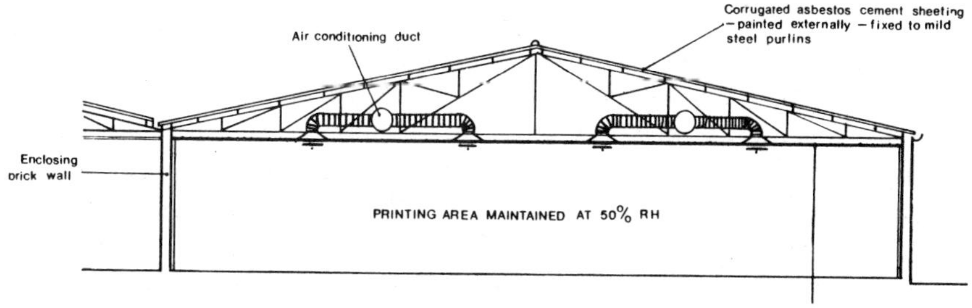

Figure 5.2 *Section through colour printing works*

A firm of colour printers wished to stabilise environmental conditions in their works and appointed specialists to advise them. The consultants recommended conversion of the existing ventilation system to full air conditioning to maintain a relative humidity of 55% throughout the year with internal temperatures of 24°C in peak summer and 18°C in winter. They also recommended insulating the roof. The printing works had overall dimensions of 70m × 23m; there was an asbestos-clad pitched roof on steel stanchions with brick walls, and a false ceiling under the roof trusses comprising translucent PVC panels in a steel T section grid [Figure 5.2]. Initially the consultants advised roof insulation to be applied under the asbestos cladding of the sloping roof, but they later said it would be more beneficial to insulate the false ceiling. This was done at the same time as the air conditioning, the PVC panels being replaced by mineral fibreboard infill panels overlaid with 50mm fibreglass quilt. Severe condensation was experienced at the asbestos roof, the suspended ceiling was saturated, production was affected, and extensive remedial work had to be carried out including replacement of the entire suspended ceiling. This would have been avoided if the consultants had stuck with their initial advice for a warm roof, as the trouble was due to moist air migrating from the humid working area to the cold roof void unhindered by any vapour barrier at the false ceiling.

Compatibility with building

The design of the air conditioning system must be in keeping with the building and its location, and conversely the building design must incorporate all requirements imposed by the selected method of air conditioning. Although a self-evident truism, it requires full and conscious communication between the air conditioning designer and all other people concerned with the design to make it happen.

Often this communication is lacking. An obtrusive condensing unit was installed in a prominent position on a listed building without checking that planning consent had been obtained. A fully glazed building had been designed with a particular solar control glass, but without reference to the air conditioning designer the glass was changed to a grade with much higher thermal transmittance. Insulation was omitted from the roof of an air conditioned supermarket because the quantity surveyor assumed it had been allowed in the M & E contract although the engineer's drawings showed the insulation to be by others.

Every designer must ensure that his design assumptions are known and implemented by all concerned with the project.

The design team on an extension to a public building chose a dual duct air conditioning system with a blow-through air handling plant in a builders work plant chamber subject to the system operating pressure of 2000 Pa. The plant could not achieve its specified performance, which proved to be due to air leakage through the unlined walls and roof of the plant housing. The leaks could not satisfactorily be sealed, and eventually the plantroom was rebuilt with metal linings. The failure occurred through breakdown in communication between architect and consulting engineer. The engineer's design was based on a 3% leakage rate, being an arbitrary three times the maximum permitted leakage of the current HVCA specification for sheet metal ductwork. No leakage rate was specified and no-one was told of the design basis as the engineer assumed the architect would know the plantroom had to be airtight. For his part the architect made no special provision as he expected to be informed of any particular requirements for the builders work plant chambers. So nothing was done and the plant failed to perform.

An eight storey office building was being refurbished with air conditioning by fan-coil perimeter units and a separate fresh air system for the interior. The consulting engineer prepared a performance specification describing this arrangement and specifying that air flow as to be based on 9.5 l/s per person from considerations of comfort and ventilation. The consultant used this figure for assessing shaft sizes for the architect but did not check that it would be feasible. The appointed contractor found the specified air quantity to be impracticable without an excessively deep cooling coil, so he designed the system for a larger flow rate. In consequence the ducts were bigger than the consultant had stated, and insufficient space had been reserved for vertical risers. One factor affected another, and the whole air distribution system had to be redesigned with plant rooms and riser shafts in entirely different positions from those originally envisaged.

Detailed design

Having established a conceptional scheme for air conditioning that is both appropriate to the building and meets the requirements of the user and client, and having also ensured that the building incorporates everything necessary for the satisfactory operation of the air conditioning system, detailed design can proceed within the agreed parameters.

The detailed design will follow standard textbook procedures. An obvious folly to avoid is the slavish adoption of general guidelines and procedures without appreciating their range of application and limitations. Always ensure that essential features or assumptions implicit in the design are clearly stated in the specification and made an unambiguous contractual obligation. Be alert for the unusual job outside the norm where commonly used practice is no longer appropriate.

The configuration of a large public building necessitated a lot of small distribution ducts with unavoidably long runs, resulting in higher pressures in both supply and extract than is usual with how velocity systems and in an abnormally large area of duct surface for the air quantity. These points were not appreciated at the time and ductwork was installed to standard low velocity specification. Distribution proved to be unsatisfactory as air leakage was greatly in excess of the 10% allowed for in the design, and all joints had to be sealed and ducts tested for leakage which entailed removing lagging and subsequent reinstatement at considerable cost. Some extract systems with long runs of small ducts were operating at such high negative pressures that the ducts collapsed, and to prevent further incidents relief devices had to be installed in all potentially vulnerable plants. The defects could have been avoided if the designers had appreciated the implications of the unusually long duct runs, had told the contractor what assumptions had been made as to duct leakage so he had a limit to work to, and had realised that the high duct pressures would exceed the limits of the duct construction.

A common design fault is the failure to think through the operation of the plant in all its modes, often after allowing for some special feature without going on to consider the effect on general plant operation. In one case, a failsafe system was devised to protect stock in a heat-producing process by opening dampers to give natural ventilation on electrical failure, but it proved useless when an emergency occurred as the designer had not ensured that circulating fans were held off until the dampers were reset on restoration of power, so stock was ruined with heat build-up from short-circuiting.

Even when there are no special requirements or unusual features, the designer does not always check through his design as he should to ensure that each component is mutually compatible.

An office building had four air conditioning plants, each comprising a common air handling unit with terminal reheat units supplying the perimeter and VAV units serving the core. The supply fans were controlled to maintain constant supply pressure. The terminal reheat units had integral volume controllers and the VAV units had pneumatically operated volume controllers. The design performance could not be achieved although each component complied with specification. Investigation showed that the terminal reheat units were incompatible with the VAV units (although from the same manufacturer) as the self-acting volume controllers of the reheat units necessitated a higher duct pressure than the VAV units could cope with. If the designer had followed through the practical operation of the system he would have realised that the fluctuations of pressure and volume would exceed the range which the specified equipment could handle.

Installation

Good practice during installation is aimed at ensuring that the work complies with specified requirements; that matters left to the discretion of the installer are dealt with competently; and that workmanship is to an appropriate standard. This requires experienced and diligent operatives and good supervision by both contractor and client or consultant. Many cases of poor workmanship that are not picked up and rectified at the proper time through inadequate supervision cause added difficulty and expense when they eventually come to light after the job is completed. Common faults in workmanship are badly made pipe joints, ineffective painting, inadequate lagging, imperfect vapour sealing and untidy wiring. Work subsequently hidden from view demands exceptional vigilance in supervision. Often poor accessibility occurs through lack of foresight, which at best impedes maintenance and in extreme cases can lead to severe damage if the effect is to hinder emergency action.

An office building in central London utilised the ceiling void for air distribution with air inlet through perforated acoustic tiles. Most of the tile perforations were blind, but some were drilled through to act as air passages. The effectiveness of the system depended on maintaining the right pressure in the ceiling void in relation to the proportion of perforations. The installation could not be made to work effectively although each individual component appeared to be correct. Eventually a survey of the ceiling void showed that while the suspended ceiling was horizontal the soffit of the structural slab above was not, and in places the depth of ceiling void was 100mm less than it ought to have been. Remedial measures were taken to counteract the effect of the incorrect slab levels, but the system still did not work properly. Air volumes were down throughout the building, and it was decided to abandon ceiling tile distribution and install diffusers instead. Only later was it discovered that incorrect ceiling tiles had been supplied by the manufacturer. The system had been designed for tiles having 2% perforations; tiles were ordered on this basis, and the delivery note confirmed it; but when a tile was belatedly taken down and measured it was found to have only 1% perforation. This halving of the free area coupled with the pressure losses due to the reduced ceiling void height had caused the substandard performance of the air conditioning system. The problems had to a large extent been brought about by divided responsibility — the services sub-contractor was responsible for the air conditioning but not the ceiling tiles, which were supplied and installed by a separate sub-contractor, while it was the main contractor who constructed the sagging floor slabs. With effective coordination and supervision the ventilated ceiling could have worked effectively.

Commissioning, testing and handover

This most important phase of the project is the time to check that everything is working as the designer intended, that the separate components are properly integrated, and to make fine adjustments to the plant to obtain the desired operating characteristics. This is also when the client takes over responsibility for operation and maintenance of the plant. There is no shortage of good advice on how it all should be done, notably in CIBSE Commissioning Codes and BSRIA Application Guides, but how often are they properly implemented? How many jobs are handed over with only makeshift record drawings and operating instructions, if indeed there are any at all? Perhaps less evident are the number of jobs where we think the correct procedures are being scrupulously observed until close examination shows they are not.

Comprehensive testing was undertaken by a specialist organisation on the air conditioning system in a multi-storey office to ascertain plant performance. Air flow rates had been measured in all main and branch ducts, and the traverse readings were recorded in the test report. For the first air conditioning plant tested, air flow rates were also measured at individual outlet diffusers using a calibrated hood. As to be expected, the two sets of test measurements for this plant using different instruments in different locations did not give identical results although they were sufficiently close to be credible. What was surprising was that on the six other plants tested the total of air flow reported to have been measured at individual outlets agreed exactly with the air quantity measured in the main ducts. Further scrutiny indicated that the reported figures for individual diffusers could not have been actual measurements as they purported to be. Other parts of the test report showed omissions and inconsistencies, and the whole report covering air flow rates, water circuits and acoustic readings had to be rejected as unreliable.

During an investigation on a malfunctioning air conditioning plant the contractor was adamant that his pre-operational checks were so thorough that the fan could not possibly have been started with dampers inadvertently left closed. The completed system was first checked by the erection team; then checked by the site engineer who verified it complied with specification; and then it was inspected by the commissioning team who certified it was ready to start. Not until all these procedures had been completed and signed for was the fan allowed to be turned over, so anything wrong with the installation would have had to escape the notice of three separate teams making checks over a period of several weeks. That was the principle, and the contractor obviously believed it worked. But examination of the commissioning sheets showed that all the separate checks had in fact been signed for by the same man on the same day. Far from being an impossibility for the fan to have been started up against closed dampers as the contractor initially stated, it turned out to be the most likely explanation of the malfunction under investigation.

Always check everything and take nothing for granted. A purpose-built centrifugal refrigeration machine could not get up to starting speed before cutting out on low oil pressure. The compressor, motor, starter, and electrical supply were checked and found to be satisfactory. The starting torque given by the US manufacturers of the compressor matched the ratings of the motor and starter. Every component appeared to be working exactly as it should, and yet the compressor could not get up to speed quickly enough. The trouble was finally traced to the compressor manufacturer's UK technical data for 50Hz supply. In converting from 60Hz supply the starting torque had been taken as directly proportional to the speed ratio rather than the square of the speed ratio as it should be for a geared machine. Hence the published technical data for 50Hz machines were wrong, and the motor and starter manufacturers had been given incorrect information.

Another example occurred when an air conditioned sports complex had problems with their water chilling plant. After several faults had been put right the system should have worked satisfactorily but it still could not achieve the load. The condensers appeared to be at fault, but they had recently been retubed and should have been working well — indeed the heat transfer specialist who did the retubing said the performance would be enhanced as the patent type of turbulator used had exceptionally good heat transfer characteristics. This proved to be the clue, as investigation showed the replacement turbulators were not intended for refrigeration use and had a much higher resistance than the standard type. When the correct turbulators were obtained from the refrigeration manufacturer and fitted in situ, the plant started to perform as it should.

General lessons

Recurrent themes crop up time after time with air conditioning systems that are defective or not operating as well as they should. Lessons to be learned are:

- Ensure there is sufficient air where it is needed
- Guard against air leakage from both ducts and building fabric
- Take account of the whole operating cycle
- Beware condensation — ensure adequate vapour sealing in the right place
- Make sure everyone knows the requirements
- Make sure everyone knows the limitations of the scheme
- Keep within limitations of design procedures and assumptions
- Avoid unnecessary complication
- Take nothing for granted
- If costs are squeezed, reduce facilities rather than standards
- Provide adequate supervision, particularly on work to be concealed
- Ensure thorough commissioning and formal handing over
- Do not divide responsibility

Conclusion

Each of the case histories quoted has been simplified to demonstrate a particular point. In reality most practical situations are more complex, and I conclude with a final study which illustrates that problems on malfunctioning jobs seldom have a single cause.

A financial institution appointed architects for a new head office with a brief for a simple building without ostentation. The architects recommended air conditioning as the site was in a city centre and engaged consulting engineers, who prepared a performance specification and invited proposals from designer/contractors. The accepted scheme was for heat recovery air conditioning. Work started on site but was delayed through adverse weather and other causes. Just before scheduled completion the .client company merged with another, resulting in extensive planning alterations affecting senior management and directors' suites. After occupation there were persistent complaints of inadequate heating and poor air conditioning throughout the building. Many defects were found, including cracked boiler sections, leaking windows, and air infiltration through the structure. These faults were dealt with as contractual defects, and attempts were also made to improve conditions by enlarging the boilers, installing extra heaters, sealing windows, and rebalancing, but to no avail as the complaints continued. Without notifying the design team, the client called in another consulting engineer to report on the system. The new consultant recommended extensive alterations which entailed scrapping the heat recovery and replacing by conventional air conditioning, putting in larger boilers, and substantially altering much of the air distribution. These recommendations were initiated while the original consultants were still endeavouring to get the work put right under the contract. The new consultants were not told of the original brief, and their advice was directed towards the currently stated requirements of the present clients. The advice was also open to criticism in respect of the test figures and technical calculations on which it was based. The result was a very expensive replacement of original plant, much of which was unnecessary to remedy the genuine defects but served to enhance the system to the more grandiose expectations of the new clients. The real problems had been a combination of poor design, unrealistic assessment of heat recovery performance under operating conditions, insufficient means of balance and control, ineffective supervision, and a badly constructed building, exacerbated by last minute alterations following a change of client. Everyone concerned was at fault to some extent, including the design team, the building contractor, the air conditioning sub-contractor in both design and contracting roles, the client, and not least the new consultant who took the easy and vastly expensive option of changing the entire system. Nobody emerged with credit, and this project provides a salutory example of how a project should not be run.

6 Software for Air Conditioning Load Analysis and Design

Jeffrey Wix and Colin McLelland

Introduction

A principal requirement of the building services engineer is the design of systems which ensure that a space can be maintained within a specified range of conditions at an economic cost. The cost in this case being both the capital cost of installation and continuing energy and maintenance costs. There are usually many ways in which capital and running costs can be balanced to provide a solution for a particular building, the best of which may be referred to as the optimum design. Knowing which is the optimum design implies that the engineer also knows which is not by having tested a number of possible solutions. Often the time is not available for complete analysis of all options: decisions may need to be based on assumptions and experience rather than on tested facts. To simplify the mathematical analysis would be an unacceptable and retrograde step whilst to employ more engineers for this purpose would not be economically feasible. The way out of this situation is often seen as the employment of a computer, a device which may not demand a salary but still needs care and attention.

To say that use of a computer is the way to approach or achieve optimum design is too simplistic as there are a number of factors which need to be looked at. This paper attempts to consider some of these as they apply to air conditioning load analysis and design. It does not go into great detail on the fundamentals of such analyses, since these factors are dealt with in other chapters.

To air condition or not to air condition

There may be good reasons why a designer must choose to air condition a building long before the load analysis phase is reached which may include the following:

- Noise in the external environment (adjacent to main traffic routes or airports).
- Close control of internal environmental conditions (certain hospital areas, precision manufacturing, special laboratories).
- Economics of trying to rent out space in a particular location (air conditioning as a requirement in high rent office space areas such as the City of London).

Until recently, decisions of this type were either automatic or based on the experience of the designer. It is now possible to conceive of computer assistance in such matters by consultation with an expert system, which is a computer program embodying the knowledge of an expert or experts. A particular example of an expert system for air conditioning is the ACE system currently under development by the Knowledge Based Systems Centre at the Polytechnic of the South Bank. This type of software is not yet available, but certainly will be in the near future. It is extremely difficult to ascertain the level of cost for this type of software since, at present, development costs remain high.

Once the load analysis phase is reached, the decision on whether or not to air condition can be based on more detailed criteria obtained by methods of analysis. Most specifically, the engineer will want to look at how the temperature in a space will vary according to external conditions when air conditioning is not present. Within the United Kingdom, the means for carrying out this analysis is the calculation of peak summertime temperatures according to methods indicated in the C.I.B.S.E. guide. Holmes (1) indicates that because this method has developed to allow manual calculation, it is not an accurate predictor of actual internal temperature and proceeds to explain how the method requires the use of more fundamental harmonic theory to become accurate. It can, however, be said that the process of decision making at this stage does not require fine accuracy, but rather a reliable indication to the engineer. For this reason, then, the C.I.B.S.E. method may be regarded as suitable.

Most software houses include within their catalogue a program or routine for the calculation of peak summertime temperature. Because of the nature of the calculation, it often co-exists with the load analysis software and within large computer systems it is unusual to see peak temperature calculation as a separate application, whilst with micro-computer systems it is frequently listed as a separate item with a typical cost of around £200.00.

Load analysis methods and their computer implementation

There are various ways in which air conditioning load may be calculated and the selection of the most appropriate may depend on a number of factors, such as:

- Familiarity with the method.
- Georgraphical location of the building under consideration.
- The need to abide by codes and standards in force.
- Preferences expressed by the client.

Typically, the methods which an engineer might contemplate include those developed by bodies such as CIBSE and ASHRAE as well as those developed by commercial organisations such as Carrier. In addition to these, software based on detailed theories of heat transfer and exchange exists.

Methods may be divided into 2 categories: those which are essentially manual methods which have been computerised for convenience (CIBSE and Carrier) and those which depend on the use of a computer to provide an answer (ASHRAE and detailed theory).

The above classification is based on the premise that computers are used solely for their numeric processing capability. A typical building requiring air conditioning will consist of a large number of rooms, each providing a significant amount of input data to a load calculation. Account is taken here of computer systems' ability to handle large volumes of data. Combining the numeric processing capability with the data handling capability means that those methods originally developed for manual calculation can benefit greatly from computerisation.

Design

The storage of data and the provision of data editing facilities within a program allows the engineer to review a number of options in load analysis and design. In terms of optimum design, and allowing for a project architect willing to be influenced by the building services engineer in matters of energy, this editing and recalculation capability can be used for a wide variety of purposes including:

- Examination of energy use as a function of building shape.
- Consideration of alternative structural configurations.
- Assessment of various shading options.

By examining the various options, and the consequent anticipated reduction in energy load, the engineer is effectively designing down both the capital and running costs for a given type of installation.

Optimisation of the building form and structure resulting from load analysis cannot be carried out in isolation. True optimisation also requires consideration of other factors which may include:

- Aesthetic impact of the building form on the local visual environment.
- Indoor air quality.
- Local planning requirements.

It should be clear to the intelligent engineer that computers do not design. It is the engineer who actually carries out that intellectual process which we term design. The computer is made to act so as to assist the engineer with the routine processes of calculation and information retrieval in the manner determined by the programmer.

Complexity of the analysis procedure

History shows us that the methods used in engineering analysis calculations have become more complex as we have striven to find ways to predict loads and performance accurately. It can be confidently predicted that the mathematics of load analysis procedure will continue to become more complex as they tend towards greater accuracy. Discussions within the CIBSE computer applications panel have included considering the recommendation of various levels of load analysis calculation to take account of different circumstances.

Whilst computer use at present makes load analysis easier by removing the drudgery of calculation, in the future the complexity of calculation required will make computer use essential.

With the growth of computer applications in load analysis, we have also seen the beginnings of a move towards simulating the performance of systems in operation. This technique is not possible without a computer and involves a detailed knowledge both of load and of the operating characteristics of system components at various load conditions.

Types of software

Software for air conditioning load analysis and design, with varying functional capability and utilising all of the calculation methods indicated above, can be found for virtually any type of computer. The following is an attempt to classify programs into various types according to the range of functions performed, the type of computer on which they can be run, and the cost of purchase. Experience shows that classifying anything in this way has problems in the grey areas between classes. It should therefore be stated that this is a personal view and does not represent the view of any corporate body or committee.

Level 1

Software which will operate on a home computer such as an unmodified BBC, Amstrad, Sinclair or MSX type. Typically this type of software will carry out a single function such as a heat gain calculation or calculation of solar angles or solar radiation at a particular location and time. Calculations are usually concerned with a single design value such as maximum load at a time specified by the operator since the processor speed is too limited for extensive calculation. Programs will predominantly use data entered at the time of program operation as opposed to data stored on files due to limitations of storage capacity and data access time. This aspect is, however, rapidly becoming a grey area as disk storage technology falls in cost to the extent that it is economically viable for this type of computer. The cost range for this type of software is tens of pounds.

The use of level 1 software must be regarded as limited and not particularly helpful to practical design. It can be useful in educating engineers where the intention is to examine the variation of a particular function. Sometimes, source code is available for this level of software and this can be useful in cases where there may be a need to vary the program code.

Level 2

This level of software carries out a single function in the same manner as the level 1 software but operates on the office personal computer such as the IBM PC, Apricot, or Olivetti M24. The functional capability of the software is extended both because of more extensive data storage capability and because of greater processor power. These are utilized in the storage of data on files which can later be retrieved, modified and rerun for the purposes of optimization or revision and in the greater range of calculation functions carried out. As an example of the latter, it is reasonable to expect a program of this type to carry out hour by hour heat gain and/or cooling load calculations. Clearly, the quantity of output data to be expected from these programs is greater than that from level 1 software and it is the responsibility of the engineer to analyse the data. Since many of the computers which run level 2 software also have some graphics facilities, many software houses have taken advantage of these to present output in graphical as well as tabular form, a provision making analysis a much easier task, particularly when searching for a maximum or minimum. It is usual to find maximum/minimum search by interpolation available in conjunction with graphics.

Availability of level 2 software is not only restricted to office personal computers, for there are many examples of this type of program available on minicomputers, although this type of hardware is now more likely to be used for higher levels of software. A typical cost level for this type of software will be in the hundreds of pounds, with £200–£500 being usual. Minicomputer implementations will be significantly more costly.

This type of software can be useful in the smaller design office and for educational purposes in a similar manner to level 1 software. Source code will be rarely available. At present, this is the most widely used level of software and includes the most popular ranges of microcomputer software including:

- Amazon
- Cymap
- Hevacomp (the Hevastar suite)

Level 3

This may be regarded as multi-function software since it endeavours to carry out a range of calculation types using the same data. An example would be a program which calculated peak summertime temperature as well as heat gains plus allowing for plant operating characteristics and calculation of energy use. This level of software can also be extended to include software which uses a common database for various calculation types such as load analysis, illumination etc. The common aspect of software at this level is the use of standard methods of calculation, most typically CIBSE or Carrier. It is the type of software most often found on the more powerful office personal computers with larger RAM memory configurations although it may also be found on minicomputers. A typical level of cost will be around £1000 for a microcomputer implementation with minicomputer versions several times more expensive.

Level 3 software can be regarded as the most useful type available for normal design purposes due to the efficient use of data input. Regrettably, there are few satisfactory implementations of a satisfactory database for building services software. In this sense, database is considered in its most general sense of relating information across multiple tables or in hierarchies rather than in the program controlled files currently used. Implementations currently available include:

- Cymap (partial implementation)
- Focus (SECAD system used by Health Authorities)
- Hevacomp (design database)

Level 4

This is the most powerful of the load analysis and design software and should be used where complex design problems are to be solved. It has all the features of lower-level software in being able to deal with single or multi-functions, plus the additional capability of being able to model the dynamic thermal behaviour of the building. Many of these programs are also being extended to carry out system simulations so that they are truly becoming design tools rather than fast calculators.

Amongst the facilities which may be expected of this type of software are:

- Energy simulation of a building against real climate data.
- User definition of building and plant operational details.
- Economic analysis based on capital and running costs.

The calculation bases on which programs are built are likely to be beyond those found in standard guides and manuals. Many have their roots in research work done at universities and other establishments, usually for the purpose of defining more accurate methods of prediction. For instance, the ESP program from ABACUS uses what is termed a first principle implicit finite difference technique, whilst the TAS program from Amazon Energy utilises the response factor technique with the imposition of additional unique techniques. Clearly, programs of this nature require a degree of validation to give the user trust in their output. Attention is drawn here to the BRE/SERC project to define validation models for such programs.

A user might also expect to see more advanced methods of data input with level 4 software. Again taking the ESP and TAS programs as examples, these use mouse driven interfaces for data entry. Given the range of facilities available with these programs, the use of a friendlier user interface can be a positive benefit.

The type of hardware for running these programs is essentially a mini-computer. The term mini in this case is used with caution since the most popular implementations now appear to be based on desktop systems such as the Apollo. An ESP implementation has been placed on the Whitechapel computer, which must truly be defined as a supermicrocomputer (super in the sense of its processing and data handling capabilities). We have just seen the release of a machine called the IBM PC RT which uses a RISC processor (Reduced Instruction Set Chip). IBM entry into this market must be seen as the seal of approval for this type of hardware technology and we must therefore expect it to become more commonplace.

A recent development has been the linking of this level of analysis and design programs with 3D computer aided design programs. For instance, the programs produced by Facet are being linked with the Intergraph CAD system.

The functional capability offered by this type of software does not come cheaply: prices are in the order of several thousands of pounds.

Selection of method

To carry out a series of analysis and design calculations based purely on one method is certainly possible. However, a reasonable question to ask is whether this is either a good idea or even the right approach. A design will develop over a number of stages from a rough preliminary level to a complete design against which tenders can be invited. There can be little reason for carrying out extensive, advanced building dynamic thermal performance calculations at a preliminary stage. It is quite possible that an approximate method, similar to the 1959 guide example indicated, is perfectly adequate. When probable performance does need to be analysed, then approximate methods are not appropriate and computer based tools come into their own.

The next question must be which computer tool? We may have available software at level 2 or 3 and software at level 4. In most cases the level 2/3 software will prove adequate, since most buildings will not require complex modelling of zones and performance. In all probability the design fee will not allow the engineer the necessary time to build and tune the model to the required extent. Of course the level 4 software can be used whenever required, but the building with complex load characteristics is its most appropriate constituency.

Education and training

It is necessary to make some comment on the education and training of engineers to work with computers in load analysis and design. Many colleges and universities now use level 2/3 software in their courses. Very few use level 4. There are no known formal courses on the use of computer techniques for the older engineer. No known course specifically attempts to train engineers in keyboard skills.

It is important that the engineer using a particular computer program should have some awareness of its technical basis, because at the end of the day the engineer is carrying out the analysis and design and not the computer. In a recent lecture, Clarke (2) considered the possibility of engineering diverging into a mathematical discipline, with a deep understanding of techniques, and an applied discipline, with the ability to turn the mathematics into a working system.

U.S. Public domain software

This chapter concludes with a brief discussion of public domain software for load analysis and design. This is the result of federal funding of development projects carried out in the United States where, under the rules for this funding, the results must be made available to the public at large. Where computer software is concerned, distribution is typically by tape, with the purchaser paying only a fee for the tape and administrative charges. Thus, it is possible to obtain the source code for some very powerful load analysis and design software for a nominal cost of a few hundred pounds. Examples of this are:

- BLAST
- DOE II
- NBSLD

The offer is probably not as good as it sounds, since such programs are effectively unsupported and the user must load the program onto the computer on which it is to operate (with all the inconvenience of making sure it operates properly). What could be useful is the extraction of source code routines from these programs in cases where software development is required.

Conclusions

Computers are no more accurate at calculating loads than the methods on which they are based. Where a method has been developed to provide greater accuracy, then a computer implementation of it will allow the engineer to gain a more accurate result. However methods which are not wholly accurate are in daily use in engineers offices and produce systems which work. It can therefore be concluded that computers can be effectively used as an indicator of load and the necessary system design to meet that load.

Software has been discussed as existing at various levels. It is our opinion that the greatest benefit to be derived from using a computer is not mimicry of the manual process with standalone programs. This type of program is useful in educational terms to explain a process and of some use in normal working. The highest level of software is that which allows dynamic analysis of building and system performance. With this type of software however, the user is required to enter a considerable volume of data and make a preliminary assessment of interactions. With some exceptions, the interface between the user and such programs is not very friendly. Therefore we conclude that, unless the project is of a highly energy sensitive nature, or the user wishes to obtain some public relations benefit, thermal modelling software is not yet a practical solution. It requires a better interface, probably to come via integration with CAD software.

Where the greatest benefit can be derived is in the use of database driven software. Whilst comments have been made about the nature of the databases used, nonetheless the facility to reduce data input and to merge information between, for instance, a lighting program and a load calculation program can be very useful. Our view is that the next few years will be marked by further development of this type of software before dynamic modelling becomes the really useful tool we all hope it will be.

References

1 Holmes M.J. Computers, Engineers and Thermal Loads *Microcomputer Newsletter No 5*. Building Services Research and Information Association. April 1983.
2 *Proc. Dynamic Thermal Modelling Workshop unpublished*. Building Services Research and Information Association. October 1985.

7 Lloyd's of London

Introduction: *C.T. Barker*

In this century Lloyd's have twice outgrown buildings that they had commissioned to accommodate the insurance market and associated facilities. In 1977, they asked six leading international architects to present a solution which would allow the insurance market to be efficiently housed, but at the same time allow for approximately 200% further expansion. The result of this competition is the new Lloyd's building, designed by Richard Rogers and Partners, with Ove Arup the engineers for the building services and structure.

Concept

The building consists of a rectangular block, containing an atrium, and surrounded by six satellite towers. A two-storey basement occupies the entire site. As a general principle, the main building above street level contains underwriting and office space, the satellites — vertical circulation both for people and building systems, and the basements — plant and service activities.

The most important single aspect of the design is the need for adaptability. By concentrating lifts, stairs, service risers, toilets etc in the satellites and supporting the building on external columns, the floor space within the columns is entirely unobstructed and restrictions on use are minimised. The Atrium, in addition to providing a source of daylight to the core, enables the insurance markets which are located around it to operate visually as one market, a requirement that has always been important to Lloyd's.

General description — building

The main building is a rectangular block, 68.4m × 46.8m. The lower ground floor is set slightly below street level and contains public areas, a restaurant, and the reinstated old library. The floor above this, together with the first four floors (galleries) around the Atrium are the Room, i.e. the insurance market. At the beginning of the design, only one floor plus one gallery constituted the Room, but market growth in the last six years has resulted in 100% additional space requirement. In the design the market can expand up to and including Gallery 6. Other floors not used for underwriting will be used for offices, with the exception of Galleries 11 and 12, which contain the Committee and Executive Suites. This includes the Committee Room, which has been adapted from the original Adam Great Room at Bowood House, Wiltshire.

In order that the single space Room can be achieved, the underwriting galleries open directly onto the Atrium, the other galleries are glazed.

The first six galleries form a complete ring around the Atrium, but above that the galleries are cut back to suit the rights of light of adjacent buildings.

Each floor division consists of a number of horizontal zones.

1 The beam grid.
2 A high level services zone which contains lighting extract air, smoke detection, sprinklers.
3 A steel panel which sits on stub columns at the beam intersections and acts as permanent form-work, supports acoustic panels, and provides a services support grid.
4 A concrete floor which provides a fire barrier and supports a computer floor.
5 The low level services zone which contains the supply air, the local heat pumps, condensing water distribution, and in underwriting areas additionally chilled water distribution, electrical power, data and telecommunications systems.
6 The raised floor which is a modular panel system, incorporating air inlet grilles and flow outlet boxes for electrics.

The zoning of the floor is fundamental to the concept of an adaptable space.

1 The services can be re-routed easily and are not constrained by vertical structure.

2 All services associated with a zone are contained within that zone and can therefore be modified without disrupting other tenants. In this context it is worth noting that Lloyd's expect up to 2,500 individual tenancy agreements in the building.
3 False ceilings are eliminated, access to services is via the 600 × 600 floor tile or removal of light fitting, which is designed to facilitate this.
4 Partitions are suspended from the bottom beam grid and can be installed, mostly without effecting other elements. Closures either for fire or acoustic reasons around the services above the beam grid are achieved by a prefabricated dry construction gasket type closure.
5 In the raised floor zone, smoke separation is achieved by dwarf walls on a 200m² grid. Tests demonstrated that no acoustic separation in the floor zone was required.

The external cladding consists of triple-glazing units arranged on a modular basis. The outer slab comprises a standard double glazed unit, incorporating reflective glass as one skin. The second pane being a specially rolled glass with concave dimples to give the glass light reflective properties at night. The third inner pane is openable for cleaning purposes and return air from the light fittings is drawn through this space, offsetting the heating and cooling load of the building resulting from the highly glazed architectural solution.

Satellite towers

There are two main types of satellite tower. Satellites 1, 3 and 5 are principle personnel circulation towers. They are essentially all the same, and consist of a lobby, four high speed external passenger lifts, a fire escape stair, prefabricated toilet module and a service riser.

The remaining three are principally fire fighting satellites and include the Section 20 requirements for enclosed fireman's lifts, dry risers etc., in addition to being the principle electrical risers for the building.

The satellites also form the support structure for the air handling plant rooms which are located on top of them, and as support structures for the vertical ductwork distribution systems which are located externally to the building and serve the floors above street level.

Basement

The light service areas for the lower part of the building, such as kitchens, toilets, maintenance are located in the upper basement, while the heating services, central refrigeration, boiler plant, thermal storage, basement air-handling, substations, diesel generation and no-break plant etc., is contained at the lowest level. In addition, the main stores are in the lower basement, and because of traffic restrictions in the surrounding narrow streets, two hydraulic vehicle lifts of 10 and 24 tonnes are provided so that delivery vehicles can be brought into the basement for unloading.

Central energy plant

To minimise building energy consumption, chilled water heat reclaim coils are located in all the building exhaust ventilation systems. Chilled water is circulated back to the central chilling plant where this low grade energy is rejected and upgraded via an auxiliary condenser into the building heating system. In this mode of operation the central chilling plant is, in effect, being used as an electrically driven heat pump.

The central plant comprises two 1500kW screw compressor machines, fitted with double bundle condensers and one 1200kW standard machine. Detailed analysis of the building thermal performance indicated that 1500kW of building cooling load was available at all times and that up to 3000kW was available from cooling and reclaim sources that could be usefully re-used for building heating.

As the maximum demand for heat by the building does not co-incide with the maximum available reclaimed heat from the chillers, the 400m^2 sprinkler storage tanks are used as thermal reservoirs. The circuitry allows the chillers to reject heat into the tanks while there is little or no load requirement in the heating circuit. Likewise it allows energy to be drawn from the storage tanks when the chilling plant is inoperative.

General

The building is unusual in that it combines a number of features which together have created a challenging and complex problem, and have resulted in a response which it is hoped will provide a satisfactory working environment for Lloyd's which satisfy their requirements for change and at the same time contribute to the architectural stock of the city.

Among the many criteria which have figured, some of the more important to the services design have been the high population densities in the market areas, the need to change from an office to an insurance function with minimum building disruption, the galleries opening directly onto the Atrium, energy conservation, the uptake of information technology in the City and the architectural demands on the engineering systems. The following sections take some of these elements and elaborate on them.

The architecture of air conditioning: *Graham Anthony*

During the last decade we have been witnessing a significant change in architectural thought. The doctrines of the Modern Movement, which have so forcibly shaped our environment for the past fifty years, are no longer considered inviolate and the rigid adherence to functionalism has been replaced by a less dogmatic attitude. The emergence of the *Post-Modernists*, with their language of allusion and metaphor, has irrevocably blurred the architectural image.

Architectural movements are, of course, complex affairs, part stylistic and part ideological; however, their development often reflects the social and artistic values of society and to this extent change must be partly attributable to society's disapprobation with modern architecture. Modern architecture, devoid of gratuitous decoration, relies on the formal expression of function to achieve modelling and scale. Failure to achieve this is best exemplified in the design of the modern office block, which with its emphasis on maximum rentable area and floor upon floor of identical space, leaves little scope for the formal expression of its components. The developers and their architects, motivated primarily by large profits, have systematically demolished many fine buildings in the inner cities and replaced them with banal, featureless blocks. The confidence of the public in the architect's ability to provide better buildings than the traditional ones he has sought to replace is understandably lacking. Hence the increasing strength of the Conservationist movement and its success in preventing much modern architecture from being built.

The new Lloyd's of London building is firmly rooted in the concepts of the Modern Movement. Born of ideas current in the 1960s, it has attempted to distinguish between those elements that have a limited life span and are therefore disposable from those which constitute the main building fabric and are thus permanent. It has been designed as a free standing office block hemmed in by the existing fabric of the City. Built of concrete, glass and stainless steel, its main rectangular form encompasses an Atrium extending some 80 metres high, culminating in a semi-circular glazed barrel vault. Six satellite service towers positioned around the periphery of the site are joined to the main building by glazed links. Like much of Richard Roger's work, it invokes a passionate response, mixed, in about equal measure, between enthusiastic acclaim and critical dislike.

The architectural philosophy behind Lloyd's is essentially one of Expressionism. The aim, to provide legibility of the component parts, thereby establishing their inter-relationships, and to juxtaposition them so that they form a sculptural entity. The overlayering of elements and the attention to detail of each component becomes the method by which the grain and scale of the building is defined. Historically, in classical architecture for instance, this would have been achieved by the use of motifs within the forms of the cornice, pediment and pilaster.

In the design of this building the formal expression of the service elements is as important as that of the more traditional ones, such as structure and glazing. Consequently, they become part of the initial design equation and not relegated to a secondary role — to be positioned where possible after the major decisions of structure and envelope have been made.

If all services are to be exposed, it follows that they need to be co-ordinated and positioned with the same degree of care as the rest of the building. All decisions, whether large or small are therefore made from both an aesthetic as well as a functional standpoint and often only after extensive drawing and modelling has been done. This inevitably increases the design, detailing and drawing programmes. To achieve this it is necessary to strengthen the relationship between the members of the design team, particularly the services consultant and the architect. It changes established roles and requires new approaches from both sides towards their respective disciplines. Compromise is inevitable. However, both parties have to have a much more flexible attitude towards solutions, sometimes abandoning traditional roles for the more innovative. The base line, is of course, the point at which each discipline knows he has taken the other to the limit. To fully understand this game of bluff and counter-bluff, knowing that the function has not been seriously impaired by considerations of aesthetics and vice versa, it is necessary for both parties to involve themselves in the understanding of each other's work and decisions have to be reached in the full knowledge of this higher order of things.

In the same way as structure can be formally expressed allowing an appreciation of the way the forces are structurally resolved, so the movement of large volumes of air in an air conditioning system can be expressed by exposing ductwork with the same degree of clarity as one might expose structural beams and columns. It therefore becomes possible with only a very limited knowledge of services to comprehend the basic rationale behind the heating and ventilation distribution system of this building by observing the various services elements and the ductwork configuration.

The six satellite towers house the main forms of vertical distribution, both for people and services. Four of these terminate in three storey plant rooms, each floor delineated by a break in the external cladding. They serve Gallery 1 upwards and their size is conditioned by the need for separate plant to service tenanted and underwriting floors as well as provision for Lloyd's to expand their underwriting activities up to Gallery 6 in the future. The remainder of the air handling units serving the two basement levels, lower ground floor and the main underwriting space known as the *Room* are situated in the lower basement, together with the major items of plant such as the chillers, boilers and generators etc.

From each of the satellite plant rooms the supply (one for underwriting and one for tenanted office) and one extract duct emerge to bend and descend vertically at the end of the lift lobby. For those familiar with the building there is a certain irony here. The glazing to the main building is predominantly obscure except for clear vision panels at either standing or sitting height depending on floor usage. The ends of the lift lobbies, are however, totally glazed with toughened clear glass, in some instances offering exceptional views over the City. It is, of course, beyond this glazing that the three large ducts descend, albeit reducing in size as they feed the lower galleries. At each gallery level a supply and extract branch duct passes through the lift lobby to re-emerge externally in a complex interchange and distribute horizontally along the facade, with stab-offs periodically occurring along the run to penetrate the external cladding at raised floor level. In detailing terms it was necessary to design a language of support structures for the ductwork to take into account the permutations of size and configuration. In conjunction with this, the actual construction of the ductwork, insulation and overcladding had to be considered. It was decided at an early stage not to pursue the concept of a waterproof cladding system similar to that of *Centre Pompidou* but to provide stainless steel overcladding which offered protection from water but was not intended to be either water or vapour proof. The eventual specification was Rockwool insulation sealed with a polymeric vapour seal. In addition a sacrificial layer of foilbacked Rockwool was used to prevent the rivets securing the rigidised stainless steel overcladding from penetrating the vapour barrier.

One of the implications of designing an external ductwork system was that it became a structural element which had to satisfy the District Surveyor. There are no Codes of Practice covering this aspect of the design and there was little data available on the structural performance of very thin tubes; none was found for spiral wound tubes. Before structural design started, load tests were carried out on some sample sections of ductwork in order to check the model of behaviour. The final designs were validated by constructing full size mock-ups of a number of typical conditions and loading them to working load, design load with its factor of safety and ultimately destruction.

Although not apparent from the road, the external services at roof level are both complex and dramatic. At Gallery 13, the basic rectangular shape of the building has been cut back by terracing at Galleries 7 and 10 to leave a U-shaped roof around the glazed atrium which projects a further storey before terminating in a barrel vault.

The basic design of the roofscape revolved around the organisation of two main elements. Firstly, the Cooling Towers and secondly the Atrium ductwork. The latter pulling large volumes of warm air from the upper reaches of the atrium and passing it through heat exchangers before discharging it to the atmosphere.

The position of the Cooling Towers was primarily governed by structural considerations as their weight had to be transmitted to the perimeter and atrium columns. Depending on the size and complexity of their supporting structure, there was some degree of flexibility in positioning them. The main concern, however, was the visual effect of the towers through the glazed north end of the Atrium. To lessen their apparent solidity, it was decided to opt for a circular form rather than the conventional square one. Three forced draught towers gave the best ratio of height to diameter and these were placed in line at the end of the Atrium.

Each extract duct positioned at high level on either side of the Atrium required a fan and heat exchanger, which, due to their weight, had to be supported from the roof slab. The ductwork, however, nearly three metres in diameter and supported from the Atrium columns on steel cantilevered brackets, needed to turn back on itself in order to connect up with the fan transformation piece. This visual effect of these large ducts projecting beyond the Atrium like vast trombones is astonishing. To achieve this it was necessary for the ductwork bend to be cantilevered as well, the joint to the transformation piece being a flexible connection and therefore non load-bearing.

Figure 7.1 *External horizontal ductwork feeding raised floors*

Figure 7.2 *Vertical ductwork descending from plantrooms*

Figure 7.3 *View of Lloyd's redevelopment*

Figure 7.4 *Design sketch of roofscape*

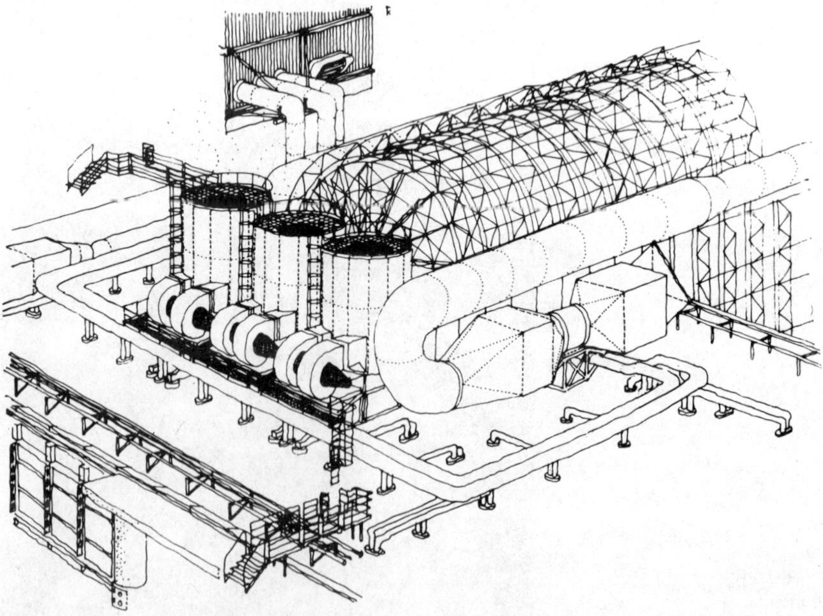

Primarily, because of the visual impact of the ductwork, it is perhaps possible to overlook the fact that the air conditioning is only a part of a fully integrated exposed servicing approach, which includes pipework and electrics, both internally and externally. This in turn must be seen within the context of the expression of all other major elements from staircases, toilet capsules, wall climber lifts to cleaning cradles at roof level. Thus it can be seen that the expression of air conditioning cannot be separated from that of the other elements but must, therefore, be part of the overall approach to architecture.

Prediction of the environment inside a 13 storey atrium: *Richard Waters*

The central area of the Lloyd's building forms an atrium, thirteen storeys high and 11 × 33m in plan. The lower three galleries of the building and the ground level form the insurance market area and are open directly on to the atrium. The remaining galleries are separated from the atrium by glass. The barred roof and the south elevation of the atrium from the seventh to twelfth galleries are completely glazed and in contact with the outside environment. A schematic view of the atrium is shown in Figure 7.5.
We were concerned that on a winter's day cold downdraughts, from the high level glazing in contact with the outside environment would cause discomfort to staff working on the atrium floor. Hand calculations and then more complicated computer analyses were used to make approximate assessments of the effects of cold downdraughts. The results from these calculations and the conclusions drawn from them are reported in this section.

Hand calculations

For a cold downdraught inside the atrium to occur two conditions must be met. The air must come into contact with cold surfaces and there must be enough power available to drive air re-circulation within the atrium. The first condition is satisfied by the large area of glass at the top of the atrium in contact with the cold air outside the building. Our calculations indicate that on a cold winter's day around 200W are available to drive the circulation of air at high level. This power estimate assumes that 0.1% of the energy loss at the top of the atrium is available to drive air

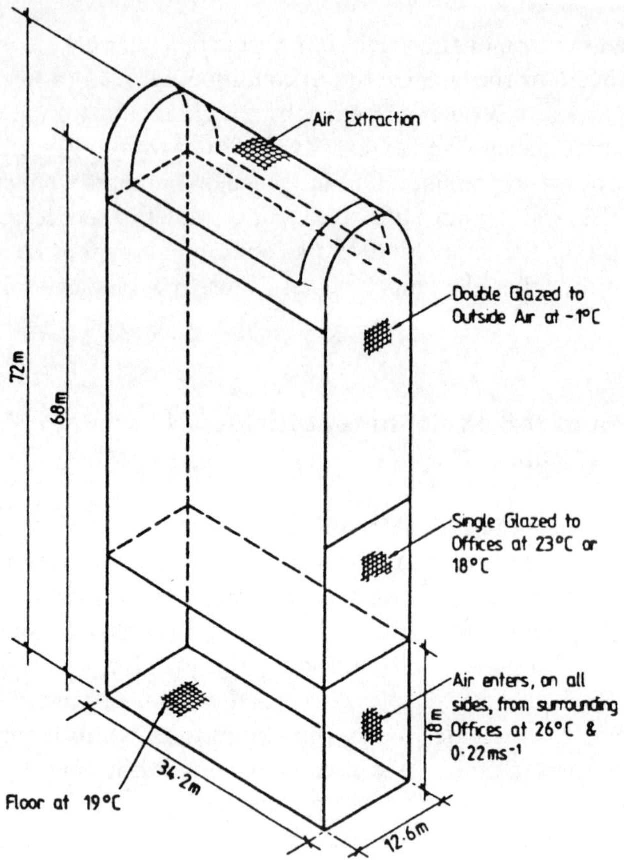

Figure 7.5 *Isometric view of atrium*

circulation — the conversion factor was found in Tritton's (1) book on Physical Fluid Dynamics.

The atrium's external glazing stops at the seventh floor of the building. There are then three sealed floors, galleries 6 to 4. Our calculations indicate heat conducted into the atrium space from these levels will slow but not stop downdraughts.

The bottom three galleries plus the ground level of the building open directly onto the atrium. This area will become the underwriting room. We estimate that the air from these levels has a speed of about 0.5 to 1.0 m/s as it rounds the corner from the ceiling of the underwriting room into the atrium. It is likely that this will be sufficient to break up the cold downdraught.

The peak motion of air above the seventh storey in the atrium was estimated from the thickness and velocity relationships for buoyancy driven flow over a flat plate. Velocity (V) and thickness relationships of the type:

$$\delta = x\ (Gr)^{-m}$$

$$V = \frac{v}{x}\ (Gr)^{n}$$

were used with m and n taking the values prescribed by Tritton (1). Assuming temperature differences of the order 10–20°C between the glass surface temperature of mean internal air temperature it was estimated that the maximum air velocity would be some 1.5m/s; and that a typical length scale for the width of fluid would be approximately 2–3m.

Computer analyses

The purpose of these analyses was to re-examine the air flow inside the atrium with a different type of calculation technique, and thus qualitatively check the values and effects predicted by our hand calculations.

It was decided to use a fluid dynamics field modelling technique to understand the effects of cold downdraughts within the atrium. A literature survey showed that only a few field modelling techniques were capable of simulating flows at the high Rayleigh Numbers found in this atrium. A field modelling technique produced by Prof D.B. Spalding (2) selected for this study. One of the major reasons for this selection was the work by Markatos and Pericleous (3) subsequently published, on laminar and turbulent convection in an enclosed cavity. These workers obtained converged results at Rayleigh Numbers ranging from 10^3–10^{16}. The predicted results were in good agreement with well known correlations up to Rayleigh numbers of 10^8, and appeared to extend the validity of these correlation up to Ra of 10^{13}.

For a two dimensional flow situation the fluid dynamic equations are:

1 *Continuity* $$\frac{\delta(pV)}{\delta y} + \frac{\delta(pW)}{\delta z} = 0$$

2 *Horizontal-direction momentum* $$\frac{\delta(pVW)}{\delta z} + \frac{\delta(pVV)}{\delta y} = \frac{\delta}{\delta z}\left(\mu\frac{\delta V}{\delta y}\right) + \frac{\delta}{\delta y}\left(\mu\frac{\delta V}{\delta y}\right) + S_v$$

3 *Vertical-direction momentum* $\frac{\delta(pWW)}{\delta z} + \frac{\delta(pVW)}{\delta y} = \frac{\delta}{\delta z}\left(\mu \frac{\delta W}{\delta y}\right)$

$+ \frac{\delta}{\delta y}\left(\mu \frac{\delta W}{\delta y}\right) + S_w$

4 *Scalars* $\frac{\delta(pW\phi)}{\delta z} + \frac{\delta(pV\phi)}{\delta y} = \frac{\delta}{\delta z}\left(\Gamma_\phi \frac{\delta\phi}{\delta z}\right) + \frac{\delta}{\delta y}\left(\Gamma_\phi \frac{\delta\phi}{\delta y}\right) + S_\phi$

Where

$$S_v = \frac{\delta}{\delta z}\left(\mu \frac{\delta W}{\delta y}\right) + \frac{\delta}{\delta y}\left(\mu \frac{\delta V}{\delta y}\right) - \frac{\delta p}{\delta y}$$

$$S_w = \frac{\delta}{\delta z}\left(\mu \frac{\delta W}{\delta z}\right) + \frac{\delta}{\delta y}\left(\mu \frac{\delta V}{\delta y}\right) - \frac{\delta p}{\delta z} + pg\frac{\theta}{T_o}$$

These equations represent the steady state turbulent flow of air.

Turbulance effects were allowed for by using a two-equation turbulence model that solved for turbulent kinetic energy and dissipation. In addition to this buoyancy generated turbulence was allowed for by following the practice recommended by Rodi(4).

The field model solves the partial differential equation for continuity, momentum and turbulence in the following manner. The equations are reduced to algebraic relationships by integrating over an elemental volume and using upwind differencing to represent the convective terms. Pressure predictions are obtained from a pressure correction relationship which works on the principle that the correction is sufficient to produce velocity changes that satisfy continuity. The SIMPLEST practice was used to solve the resultant algebraic equations.

We decided to estimate air movement and temperature distribution in a series of two dimensional slices in the atrium. A three dimensional model was not used, because we were not concerned with the exact topology of the air movements, but with creating a mathematical analogy of our hand calculations.

Several simulations of slices were carried out. We varied the assumed internal glass temperatures with the intent of estimating their effect on air movement. External air temperature was taken as – 1°C. Air from the office abutting the atrium, at galleries one to three was allowed to escape into the atrium and was extracted at high level.

Sample results for a slice 34m wide are shown in Figure 7.6. The slice represents the situation corresponding to offices up to the sixth gallery and the twelfth on alternate sides of the atrium. In this simulation the results were characterised by a single re-circulation loop extending from the top of

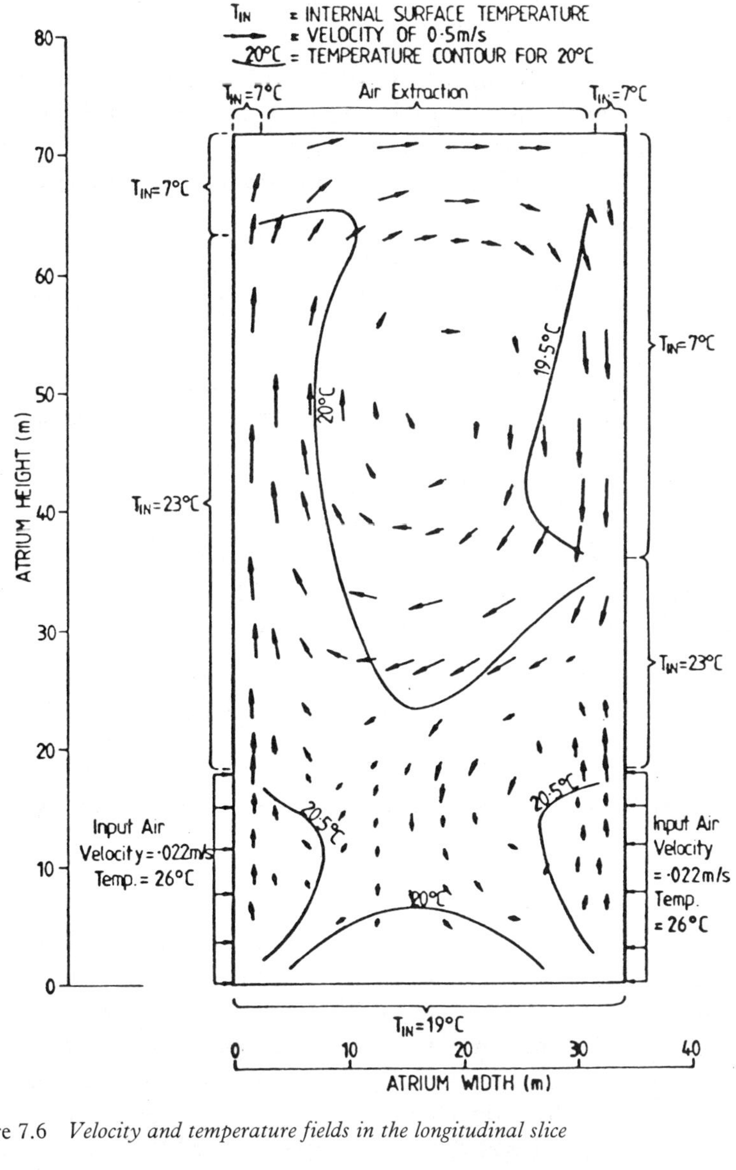

Figure 7.6 *Velocity and temperature fields in the longitudinal slice*

the atrium to about the fifth gallery. This loop has a slow moving central region; and fast moving edges where air speeds of up to 1 m/s occur. This flow pattern matches the results of our hand calculations. The computer predictions show the downdraught being predominantly destroyed at low level in the atrium by the effect of heat transfer and conditioned air from the offices abutting the atrium.

Conclusions

The conclusion of this study was that there was only a remote chance of cold downdraughts penetrating the occupied zone and creating uncomfortable conditions. To cover even this remote eventuality provision was made for the addition of deflector shields above the seventh gallery onto the atrium.

Nomenclature

Gr	=	Grashoff Number	y	=	Vertical co-ordinate
m	=	Real Number	z	=	Horizontal co-ordinate
n	=	Real Number	ν	=	Downward Velocity
p	=	Pressure	v	=	Kinematic Viscosity
S	=	Source Term	P	=	Density
V	=	Horizontal Velocity	μ	=	Effective Turbulent Viscosity
W	=	Vertical Velocity	ϕ	=	General Property
x	=	Distance	Γ_ϕ	=	Exchange coefficient for ϕ

Air conditioning using a floor plenum: *Alisdair McGregor*

Introduction

The rapid increase in the use of computer technology in the modern office has forced services engineers to reappraise air conditioning strategies for new office buildings. In the seventies computing hardware was usually located in special clean rooms which could be conditioned by a separate air conditioning system. This system could be designed to provide the correct temperature and humidity conditions for machine operation with occupant comfort criteria such as air velocities and minimum supply temperatures of secondary importance. The last five years has seen computer hardware move into the general office environment. A few years ago equipment gains of 10W/m^2 were considered adequate for most offices. Today 20W/m^2 is a common figure used for general offices and the dealing rooms currently appearing throughout the City are going above a 100W/m^2.

Reasons for choosing an underfloor supply

Early design solutions for Lloyd's included medium pressure, high velocity variable air volume, all air cooling systems serving both the core and overlapping into the perimeter areas. However, the following problems were encountered:

- The volume of plant at high level on the satellite towers.
- The size and quantity of distribution ductwork and its support system.
- The integration of air conditioning terminal equipment with the coffer and the facade.
- The projected energy consumption of the building, of the order of 1.0 GJ/m^2 which, even allowing for special Lloyd's factors of high population densities, in-house catering etc., was considered to be unacceptably high.
- High air velocities in the occupied region.

The volume of electrical and data trunking and the client's brief for flexibility of services had already demanded a raised access floor. Therefore an air conditioning system using the floor plenum was investigated. Previous work by H. Spoormaker (5,6) noted the following advantages of a floor supply/ceiling extract system.

- Elimination or reduction of ducts by the use of the floor plenum for air distribution.
- The use of heat exchange between the cool air in the plenum and the concrete structure to provide thermal storage in the structure and thus reduce peak cooling loads.
- Reduction or elimination of smoke and odour nuisance from other space occupiers through localised or workstation air supply.
- Wider effective individual temperature control range than a conventional ceiling distribution system.
- Simple system adaptation. Local high load areas can be catered for by plug in cooling units.
- If fan powered air terminals drawing air from the plenum are used then some cooling effect can be maintained if the chillers are off because of the thermal inertia of the floor slab.

By coupling the office space to the floor slab, the thermal environment of the office is very stable. This stability allows the design temperature of the occupied region of the offices to gradually increase during the day. At peak conditions the temperature floats from 19°C at 9 a.m. to 23°C at 5 p.m.

With a floor supply ceiling extract system a large proportion of the lighting convection gains do not enter the occupied region. Furthermore, the natural buoyancy of the air warmed by convective gains from people and machinery is encouraged to rise by the general upward movement of the room air. The floor to ceiling height at Lloyd's is 3m and this allows a pool of warm air to collect between 2m and 3m above floor level. The result of this is that only a proportion of the convective gains have to be cooled by the air conditioning system. Furthermore, the system becomes more stable as air can be supplied at constant temperature and differences in room heat load manifest themselves as a difference in exhaust air temperature.

Incorporation of system into Lloyd's '86 building

The principles of the underfloor system are shown in Figure 7.7. Conditioned air brought down the external ducts from the plant rooms on top of the satellite towers is pushed out into the floor plenum from header ducts. Small fan units within the plenum, known as fan air terminals (FATs), take approximately a half and half mix of primary air and recirculated room air and deliver the mixed air via flexible ducts and circular Krantz twist grilles to the room. Air is extracted at high level through the luminaires and taken along ductwork to the glazing where the warm air passes down the inner cavity of the triple glazing and is then collected in header ducts and taken back up the external ducts to the plantroom. At underwriting levels the additional volume of air required for the high occupancy density is allowed to escape into the central atrium.

The heat pumps shown at the base of the glazing are water/air reversible units. They are used to provide cooling in the perimeter zone to counteract solar gains and also morning warm up in winter. The heat pump water circuit is oversized so that additional underfloor heat pumps can be added to provide local cooling to high load zones. In underwriting levels a chilled water circuit is provided to cater for future increases in computing equipment on the floor as predicted by a report on the uptake of information technology by the insurance market.

The design gains for the underwriting levels are as follows:

- Lighting 20W/m^2
- Equipment 20W/m^2 + 30W/m^2 with chilled water circuit
- Residents 25W/m^2
- Visitors 38W/m^2.

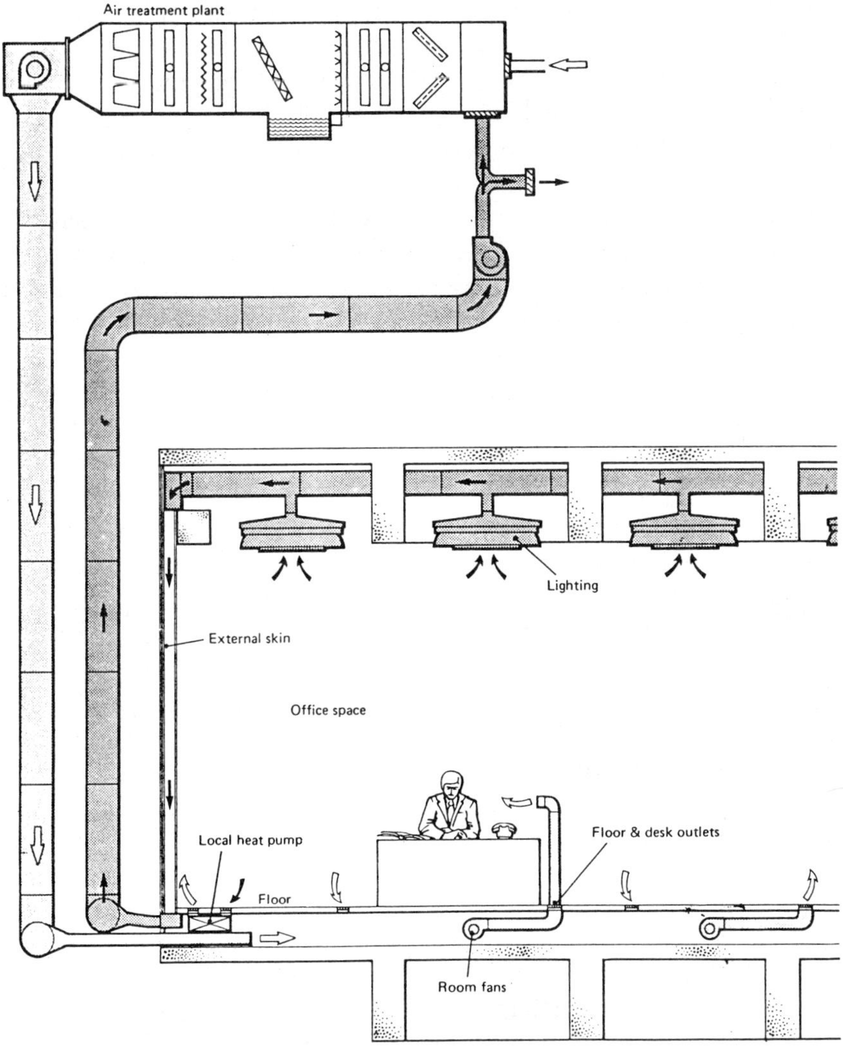

Figure 7.7 *Principles of underfloor supply to office areas*

The design gains for office areas are as follows:

- Lighting $20W/m^2$
- Equipment: 25% of floor area at $60W/m^2$
- Equipment: 55% of floor area at $10W/m^2$
- Equipment: 20% of floor area at $0W/m^2$
- People $8W/m^2$

The final design of the luminaire has resulted in a lighting gain of 15W/m^2. The occupancy gains for underwriting appear high, however, a glance into the existing room shows how densely packed it can become. To cater for these gains a supply air temperature of 14°C is required to keep primary air volumes down. This is too cold to supply at floor level hence the requirement for fan air terminals to mix room air and primary air in the floor plenum and supply air off the grille at 18°C. Exhaust temperature can exceed 30°C under full load. High exhaust temperatures cannot be achieved with ceiling supply systems as the supply air mixes with the hot air at high level before entering the occupied region.

Slab performance

Computer simulations were carried out to establish whether the concrete floor slab could be used to store "coolth" overnight and thus reduce the load on the chiller plant during the day. A finite element program was developed to enable the thermal influence of the structural slab and raised floor on the room air temperature to be evaluated.

Computer simulations were carried out to see if running supply fans at night time in summer would reduce room temperature during the day. A series of 5 consecutive hot summer days was modelled. By running the main fans for substantial periods at night, the daytime off coil temperature could be increased to 15°C. Without nighttime fans the off coil temperature needs to be 13°C.

A comparison between systems was made based on worst case conditions with an external diurnal range 16–28°C, a main fan power efficiency of 80% and a central refrigeration plant COP of 3. The results show that from an energy conservation standpoint the Lloyd's slab is of marginal benefit in peak summer conditions but could be beneficial for substantial amounts of the year.

The system is designed with an off coil temperature of 13°C so that should off peak electricity charges make nightime fan operation uneconomic, comfort conditions can still be maintained. The building automation system enables the main fans to run at 12.00 p.m. when the highest floor slab temperature is at a higher temperature than the outside air by amount set on the BAS, typically 2/3°C. As soon as the slab has cooled to 15°C then the plant shuts down. If the outside air temperature drops below a minimum then the plant shall shut down.

Feedback from the BAS when the building is operational will provide data on the economics of nighttime slab cooling.

Testing an office module at BSRIA

The Building Services Research and Information Association (BSRIA) were commissioned to carry out an experimental investigation to establish the air velocities and temperature gradients of two different systems; one employing Krantz floor grilles, the other conventional linear grilles.

The test room is shown in Figure 7.8. It consisted of a standard 2 × 2 coffer office (3.6 × 3.6m). Six tests were completed and the averages of the air velocities were less than 0.2m/s which are within CIBS recommended values.

As the room air temperature gradient was of primary interest in the investigation, the room was divided into different zones and the average air temperature was evaluated for each zone.

The following three zones were considered:

- A lower zone up to 1.5m height representing the occupied zone for a seated person.
- An upper zone between 1.5m and 2.9m.
- The ceiling-bay zone above a height of 2.9m.

The test results are only applicable to the particular circumstances of the test because there were heat flow paths in the test model that do not exist in the real building. The only satisfactory method of extrapolation from test model to real building was to construct a mathematical model. The model split the room into the three zones as above and considered all radiative and convection transfer between zones and surfaces. Table 7.1 shows the comparison of experimental and calculated results. The agreement in the case of the test using Krantz outlets is good.

Table 7.1 *Comparison of measured and predicted temperatures*

	Laboratory Temperature °C		Lower Zone Temperature °C		Extract Air Temperature °C	
	Test	Required*	Test	Predicted	Test	Predicted
E	26.5	27.5	22.0	21.8	27.7	27.7
F	27.0	26.5	22.3	22.3	27.3	27.4
G	23.8	22.5	22.2	21.5	27.2	27.4
H	23.9	24.5	22.5	22.2	27.1	26.6

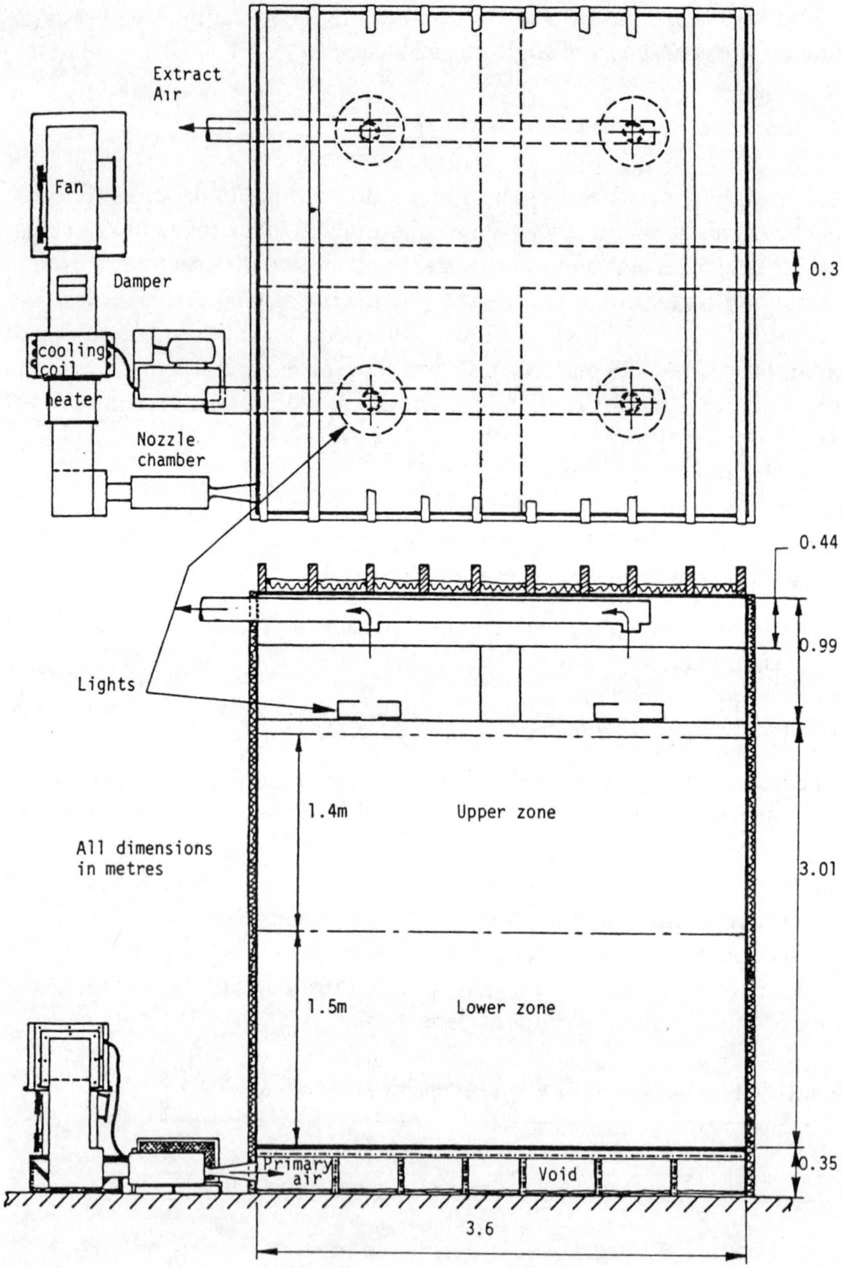

Figure 7.8 *Layout of test room at BSIRA*

The efficiency of extraction of convective heat is 60%, thus only 40% of the convective gain from the lights reaches the occupied area.

Table 7.2 presents the predicted temperature for:

Primary air	0.264m^3/s
Recirculation air	0.021m^3/s
Primary air temperature	14°C

Head loads:

Lighting level A	20W/m^2
Lighting level B	16W/m^2
Occupants and equipment	12.6W/m^2
Recirculation fan	2.5W/m^2

In order to obtain the correct trend of results, it was necessary to adjust the temperature required in the laboratory. This is to be expected, as a single measurement of laboratory temperature cannot be taken as representative of all conditions surrounding the test model.

Table 7.2 *Predicted space temperatures*

	Space temperatures °C			
Lighting Level	Supply Diffuser	Lower Zone	Upper Zone	Extract Air
A	20.9	23.0	23.8	29.6
B	20.2	22.0	22.7	27.9

The temperatures above relate to steady state conditions within the space. Thermal storage of the structure will reduce these values.

The results from BSRIA showed the floor supply system provided temperatures around 22–23°C in the occupied zone without draughts except directly above the floor grille. Smoke tests showed that good mixing took place with both Krantz and linear grilles. Temperature gradients from the test are shown in Figure 7.9.

Servicing the underwriter's box

In underwriting areas a large area of the floor is taken up by the under-

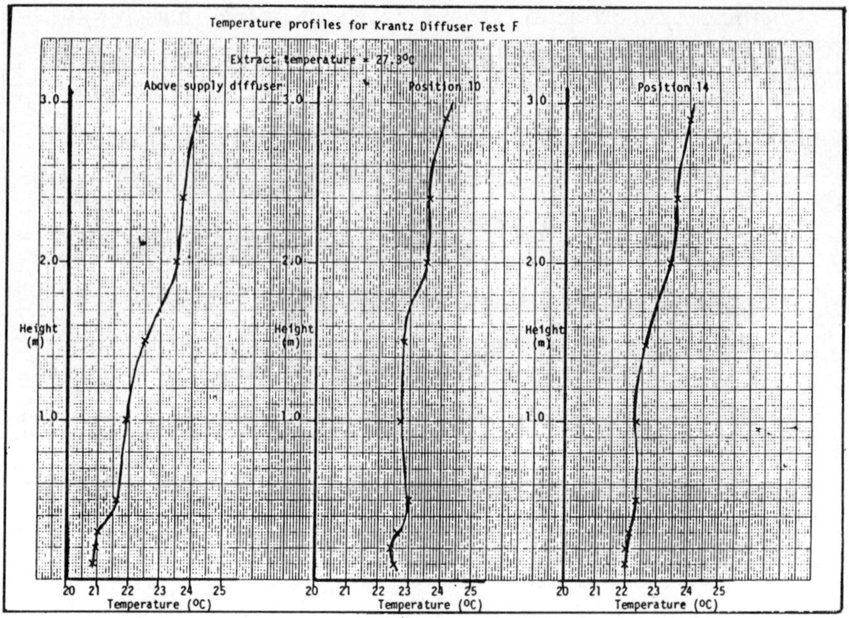

Figure 7.9 *Vertical temperature profiles*

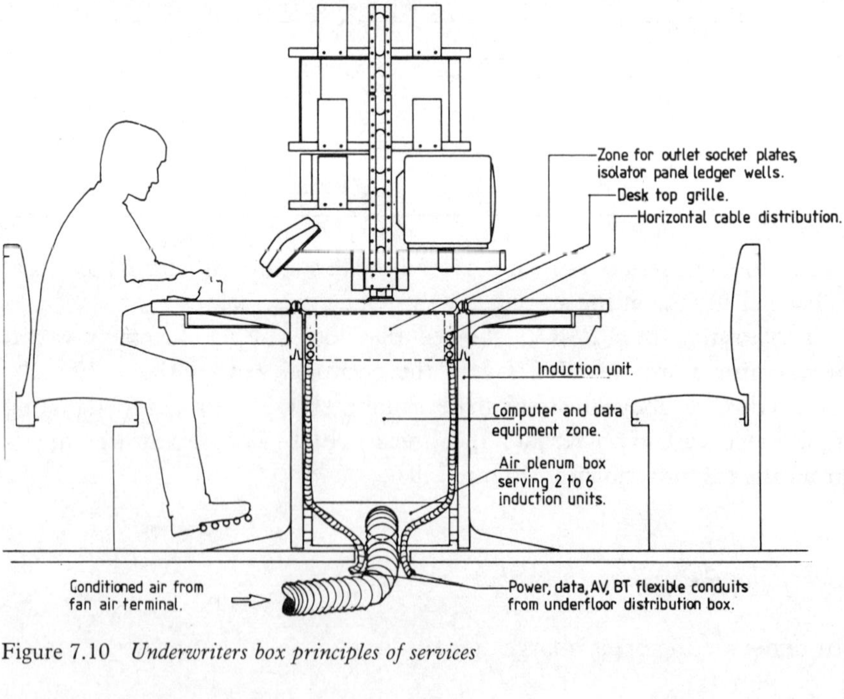

Figure 7.10 *Underwriters box principles of services*

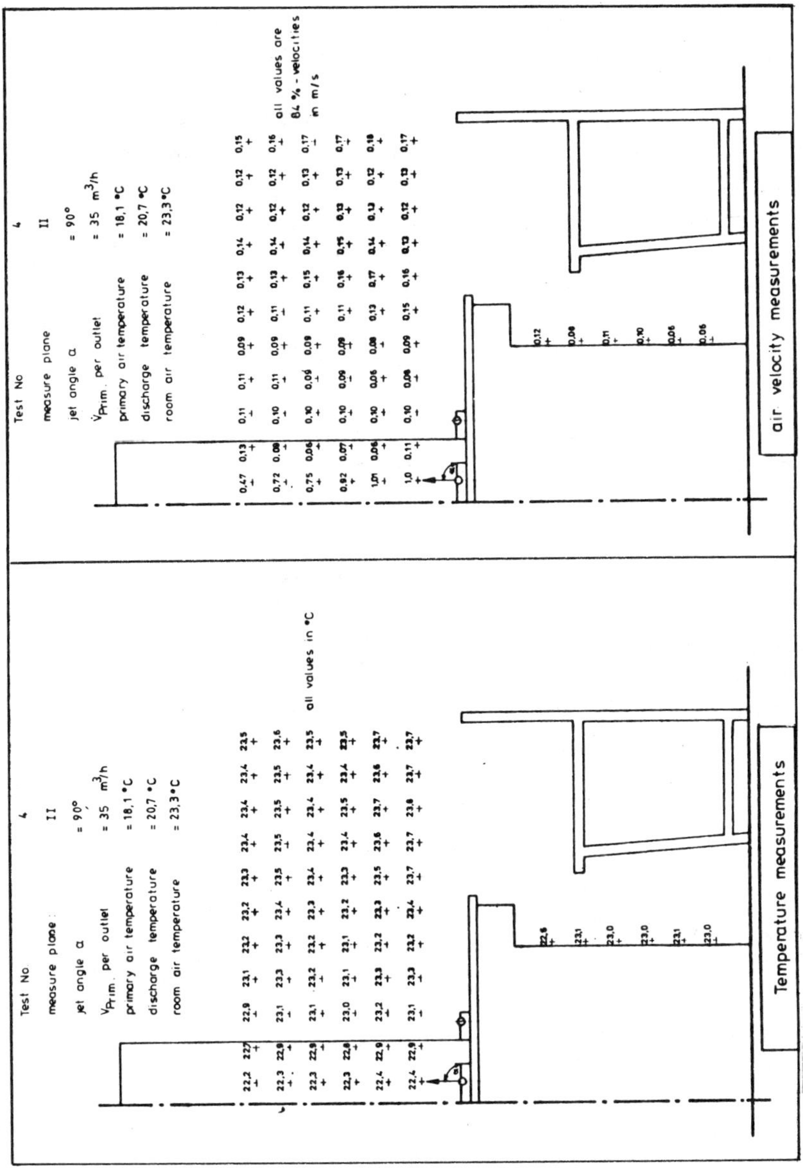

Figure 7.11 *Test results of prototype*

writer's desks or "boxes" as they are known and their seats and cupboards, leaving less floor space available for floor grilles. As most of the business takes place at the boxes, the new boxes have been designed to incorporate air conditioning, computing equipment, data, telephone, audiovisual, and power cabling. The principles of the box design plugged-in to the floor is shown in Figure 7.10.

Work by Dr Franc Sodec of Krantz has shown that if supplying air at desk level a primary air volume of $35m^3$/hr at 18°C minimum per person will provide comfort conditions. 18°C is too cold to supply at desk top level so an induction unit is provided resulting in an off grille temperature of 20°C + 0.5°C with a mean room temperature of 23°C. A prototype was tested and temperature and velocity profiles are shown in Figure 7.11.

Application of a direct digital control (DDC) system to a complex building: *Martyn Harrold*

Direct digital control is the use of software based algorithms, to achieve on/off, proportional, proportional plus integral and proportional plus integral plus derivative action control loops. Often, as in the case of the system employed on this project, the software includes all control strategies with the exception of the safety interlocks.

The Lloyds 1986 Building, due to its size, geography, complexity and the Client Brief requires a Building Management System to enable the engineer to maintain and operate the system as the designers intended.

The system is required to control:

- 45 air handling systems
- 900 perimeter heat pumps
- 1600 Fan Air Terminals
- 10,000 lighting fittings
- 7MW of heating
- 4.2MW of cooling
- 6MW of electrical input
- 3MW of standby generation

When considering the type of control system to apply to the building the options of pneumatic, electronic or DDC were considered.

The absence of terminal units from the air conditioning design eliminated any cost advantage that a pneumatic control system might have offered leaving just the fail safe very powerful actuators as an advantage over electronic/DDC control systems.

Electronic control systems are more accurate than pneumatically operated ones but neither interface neatly with a Building Management System.

Direct digital controls are more accurate than any other, they are also more flexible and far more able to cope with complex operating regimes.

Many of the problems associated with installing and commissioning a BMS/DDC system in a large complex building were expected and provision made in the contract. However some unexpected problems arose and others which were expected and for which no provision could be made in the Contract.

The first problem encountered was that of jargon and the definition and even interpretation of that definition. It is imperative that the Specification defines technical words and that this definition is the only one to be used in the Contract.

The next problem was the *what happens if* syndrome. With a conventional control system, either pneumatic or electronic the logic is defined in hardware and the sequence of action in the event of failure is very limited. Once this interlocking and control is achieved through software there are many more options available and what happens going up need not necessarily happen in reverse going down and conscious decisions have to be taken for each and every step of the logic process.

With a pneumatic or an electronic control scheme drawings are submitted for approval which show the interlocking between frost thermostats, timeswitches, fans, smoke detectors, pressure switches, firemans control switches and which show the way the controls move in the event of plant shutdown or failure. Approval of software is a major problem. The software listing for the DDC loops excluding the Building Management System functions exceed 15,000 lines of code written in a high level language unique to the supplier. Even after a two week full time training course at the manufacturer it would not be possible to approve the software and yet some checking is imperative. The only acceptable solution to this problem is to check but not approve the software listing and then at the end of the project witness each and every point being demonstrated by the contractor. To illustrate the commitment which must be made it has been estimated that this will take 250 days for two men from the Engineer/Client plus the same again for the Contractor. This has raised serious contractual problems relating to practical completion and handover. It is important to stress that all points must be checked because experience to date on other projects shows that when a contractor offers a completed BMS for acceptance normally only 50–80% of the points are correct.

One further difficulty on the subject of acceptance testing software based systems occurs when a fault is discovered in the software. A copy of the software must be downloaded from the system onto disk or tape because instances have occurred whereby previously checked and accepted parts of the system have been corrupted by the software patch. There is no way out of this dilemma as it is clearly impossible on a large system to recheck all previously checked work.

Normally on a contract with a BMS it is not essential that the BMS is up and running before the air conditioning systems and other services are commissioned.

However when the control system is combined with a BMS installation it is vital that the system is operational six months prior to practical completion to allow control valves and dampers to be moved for flushing draining and balancing. To achieve this it is necessary for the Main or Management Contractor to get the BMS control room complete, clean and secure together with its associated ventilation system completed 6–9 months prior to practical completion.

A control engineer is required to have a good understanding of air conditioning systems and plant from an electrical as well as a mechanical point of view but to date it has not been necessary to have the ability to program.

DDC Systems above all else require this ability and the industry is meeting this problem by employing programmers and unfortunately programmers do not yet have the necessary feel for Buildings and their services and it will be two or three years before this problem begins to recede.

The whole process of specifying-approval-testing and commissioning needs to be looked at carefully. The best approach is to use flow diagrams.

It needs to be said that the above description has been significantly simplified in order to highlight the salient points and avoid being bogged down in the many technical problems, conflicts and details and their solutions in the Lloyds 1986 Building.

References

1 D.J. Tritton, *Physical Fluid Dynamics,* Van Nostrand Reinhold, 1980.

2 D.B.S. Spalding, *Phoenics Manual,* Cham Limited, Bakery House, Wimbledon, London.

3 N.C. Markatos & C. Pericleous, *Laminar & Turbulent Natural Covection in an Enclosed Cavity*, Int. Jnl. Heat & Mass Transfer, Vol. 27, No. 5, p. 755–762.

4 W. Rodi, *Turbulent Buoyant Jets & Plummes*, Pergamon Press, 1982, Page 147.

5 Herman, A.F.E. & Spoormaker, H.J. — *Plenum Air Conditioning Systems*, Ove Arup & Partners, South Africa 1981.

6 Spoormaker, H.J. & McMillan, C.M. — An energy efficient flexible system for HVAC and electrical services — *Heating, Air Conditioning & Refrigeration* — March 1985.

8 Building for the Burrell Collection

Introduction: *James Briggs*

The Gallery for the Burrell Collection is owned by the People and District Council of the city of Glasgow and operated by the Museums Department of that Council. It is located in Pollok Park with access off Haggs Road. The design team of Barry Gasson, Architects, Felix J. Samuelly & Partners, Structural Consulting Engineers, the Cambridge Office of Davis Belfield & Everest, Quantity Surveyors and James R. Briggs & Associates, the Building Services Consulting Engineer were briefed by the Department of Architecture and Related Services of the Glasgow District Council.

The building

The architectural competition was won in 1972, the fundamental design completed in about 1974 and the building opened by Her Majesty The Queen in October 1983 and much has already been published (1, 2 and 3).

The total cost of the building was in the region of £17,000,000 of which the engineering services including their associated builders work cost about 50%.

The Burrell is a highly successful building in the sense it has received wide public acclaim as being both a superb place in which to display the Burrell Collection and in being a building worth going to see in its own right. It has also won a number of major awards for its architecture, as a gallery as an entity and for particular aspects of the structure and services. One of the reasons for this I am quite convinced was that the primary people in the design team struck up a working relationship which I am sure has led to a friendship that will last to the end of time, or at least to the end of our time.

The apparent simplicity has been achieved by a high level of co-ordination between architecture, structure and services. To provide the building services information for the contractual team a set of engineering services installation drawings comprising at least 400 B1 co-ordination drawings and 750 A4 details were produced, by my Practice alone, at scales ranging from 1:50 up to full size.

Details of the building and structure follow. Here I will concentrate on a brief description of the air-conditioning system.

The people

The primary members of the design team were Barry Gasson, Architects; Felix J. Samuelly and Partners, Structural Consulting Engineers; Davis Belfield and Everest, Quantity Surveyors and the Building Services Consulting Engineering Practice of which I am a Partner. The main people with whom I worked in that design team were Barry Gasson, the Architect, Frank Newby and Vic Dowsett, the Structural Engineers and Victor Bugg, the Quantity Surveyor and my own partner Peter Lawson-Smith. There is also one other person without whom all of our efforts would have come to nothing. He is Dr Richard Marks the first Keeper of the Burrell Collection in the New Burrell Collection Building.

The air conditioning system philosophy

First let me say that the building that has been constructed is virtually identical to the building I was first shown by Barry in the form of the original competition entry drawings, pictures and model.

The essential brief to our Practice was to produce a precisely controlled internal condition. We consulted with Conservation Authorities and concluded that this brief needed extension to provide sufficient flexibility to enable the actual conditions in a particular area to be set to suit the artifacts on exhibition in that area, albeit that one area connected to another with no closed doors or air locks and that there is an immense range of materials in the items of the collection.

In addition we felt it essential that the air conditioning solution had no major effect upon the essential architectural conception.

Of course, originally, we were tempted to suggest that perhaps the amount of glazing might, with advantage, be a little reduced on Barry's original ideas. For example the building could be entirely underground and that would have considerably reduced our problem. The trouble is it would not have produced a building worth going to see.

To start us off on what the air conditioning system ought to be we went over to Lisbon to see the then new Gulbenkian Museum and found the windows running with condensation — and this was in March in a sunny clime which Glasgow decidedly is not. We were also keenly aware that the nature of Gallery life can be for relatively large groups of people to stand around one object for quite considerable lengths of time while its virtues or otherwise are extolled by the Gallery Staff or visiting Lecturers.

This and the need to at least have some control of the condensation initially led us to believe that there was a need for local control of humidity. That sounded difficult but a few calculations indicated that the moisture gain from people, even at their most dense, was not likely to cause the conditions to go out of tolerance provided the humidity control at the central plant was of good quality and that the local space temperature control is reasonably precise.

Air infiltration and glazing seal

A major factor in making central humidity possible is, of course, that the infiltration of external air should be minimal. So one of the detailed matters we asked of Barry was to design the glazing to make it really air tight.

We asked manufacturers of glazing systems to tell us what the leakage rates would be for different forms of glazing and seal but none could even hazard a guess. We have therefore designed seals that we think will be effective and indeed I believe are effective. I would mention I have pursued the effectiveness of the sealing of glazing in other buildings with other manufactures and it is how possible to get at least some statement as to what the leakage should be under particular laboratory conditions.

Design basis

The primary requirement of the air conditioning system is to provide close control of temperature, generally to 19 ± 1°C and humidity to 60 ± 5%

RH. For air infiltration the design basis we took was that the sealing system of the glazing would be good enough to limit the air change by natural infiltration inwards in the perimeter of the Gallery to one air change per hour. This is equivalent to an air change overall the whole Gallery of 1/6th of an air change per hour. Since, under normal day conditions, the air pressure in the building, due to the operation of the air conditioning system, is positive, I feel sure we are achieving this and probably the infiltration into the building is much less.

This permitted humidity control to be achieved centrally for large areas and is arranged at each of 22 air plants. Temperature control was considered not to be feasible by similar central only control due to variation in local heat gains from people, lighting and the sun. Accordingly the building has been separated into some 170 zones, each with independent temperature control.

System type

The air plant areas have been selected for their commonality of building area characteristic. For example the main north exhibition area has a 3 to 4 metre wide perimeter strip along the 100 metre long 6 metre high glazed north wall, three areas with roof lights and the rest is entirely internal. Each of the these three types have separate air plants. Where the temperature change is primarily due to external influence and water leak risk is small the local temperature control is by local water to air heater and cooler batteries in constant volume air systems.

Where internal influences predominate and/or the potential damage of a water leak is high, the system is arranged as Type 2 Figure 8.1. Here the system is variable volume constant temperature from the central fan plant. This is converted to constant volume variable temperature to the air conditioned space by local recirculating fan filter units. These take an injection of air from the central plant proportional to the load. This produces an all air system that will not damage artefacts by water leakage and, being constant volume, does not produce wide temperature differences between one space point and another as occurs with fully variable volume systems. The volume variation of the central plant is arranged by pressure regulators acting upon automatic volume control dampers and inlet guide vanes are fitted so the fan absorbed energy reduces as the volume is reduced.

Another advantage produced by the system is that the air condition in the vicinity of an air stream is positively controlled in a reasonably

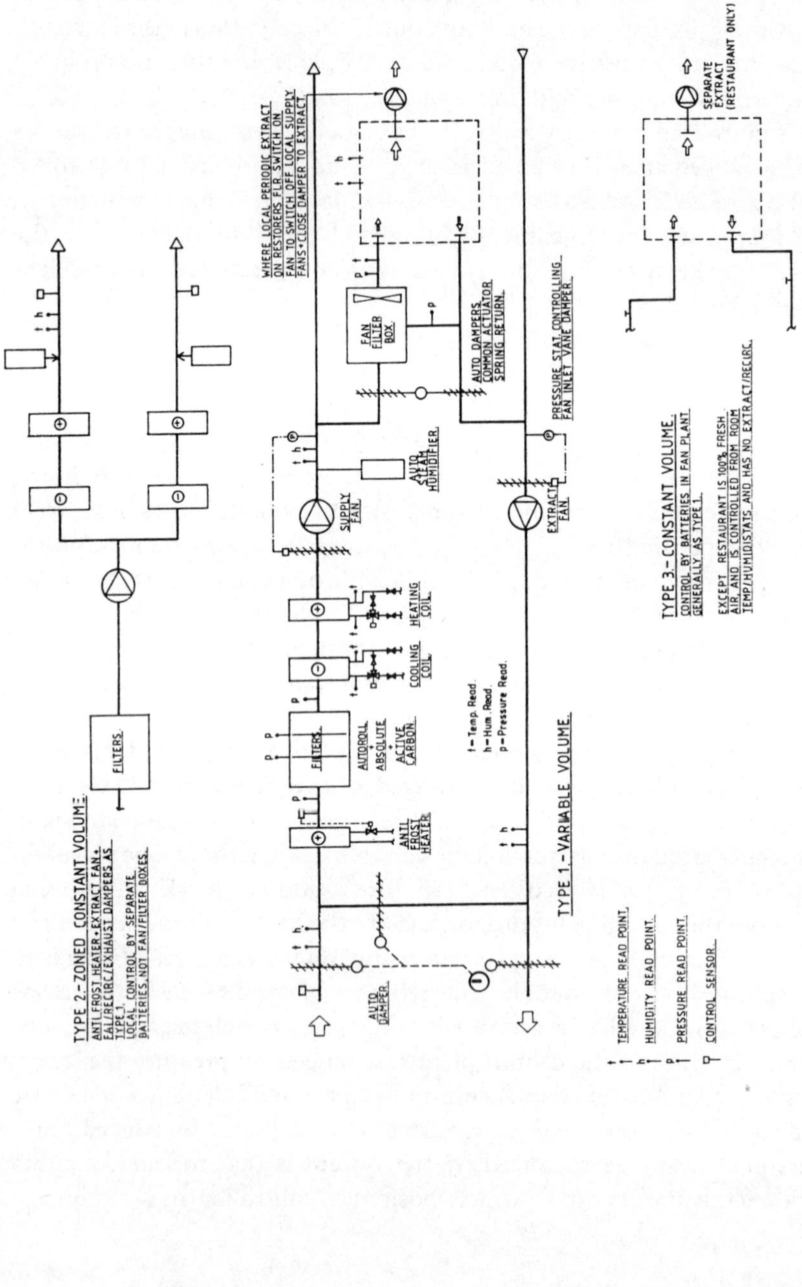

Figure 8.1 *Air system arrangement*

predictable volume of the building and the condition in the vicinity of one air stream does not have to be the same as that in the vicinity of another. Thus it is possible to control Gallery areas at air temperatures most suitable to the artefacts on display, albeit there will be a gradation from one area to another, and by careful selection of that temperature, a particular humidity may be established.

While local variation of temperature and humidity is possible we have only given facility to set the conditions differently by air plant. The actual conditions being obtained in the Gallery are almost constant between one area and another because the materials on display are necessarily grouped by type, period or some other commonality and hence a single condition has to be selected. Perhaps the Gallery may have to review this policy in the future should the cost of energy continue to rise faster than inflation.

System control

The local control of temperature has been achieved by the provision of central air plants configured for variable air volume and set to produce constant temperatures in one of two modes, cooling and heating. This constant temperature air is delivered to local recirculating fan units in variable quantities under the control of automatic dampers operated from a temperature sensor in the return air duct to each local fan unit.

To determine whether the mode should be heating or cooling there is a temperature sensor in the space independent of the recirculating fan temperature sensors and the whole control system is configured to achieve a control accuracy of less than 1/2 degree centigrade temperature tolerance. The overall permissible tolerance is 1 degree centigrade either way.

Of course the accuracy to less than 1/2 degree centigrade only applies at the sensor location and there will be variations outside the range at the sensor due to temperature gradients, air movements caused by people movements and solar radiation, where there is glazing, will cause local surface temperature rises and this too will have an effect upon the space temperature isotherms.

To minimise these effects the design temperature differential, that is the difference between the air entering the space and the required space temperature, was limited to 6 degrees centigrade. This produced air change rates of some 6.5 air changes per hour in internal areas which had no window or roof glazing, up to 22 air changes per hour for the daylit galleries which have continuous roof lights and even higher in areas with south facing glazing. This, for a 6.5 metre high space, is a lot of air and, with the local fan units introducing a constant air volume, the air distribution pattern is reasonably constant summer and winter and the supplied air mixes quickly with the space air to keep the air temperature at almost any point quite close to that at the sensor — other than in the direct path of the supply air stream.

System location diagram

A typical total system location diagram is given in Figure 8.2. This very small scale diagram is typical for each of the 22 air plants. It was produced to give the basic information of each plant on a single small sheet of paper. It was considered necessary because the main air duct co-ordination drawings were produced at 1:20. These gave the installation details needed but on many large sheets of paper that were difficult to comprehend on an individual system basis.

Air flow quantities

The air flow quantities were selected with small maximum temperature differentials in order to reduce temperature differences between different points in the conditioned space, as already mentioned, and also to achieve air flows in the required directions. The essential direction of air flow needed to be out of areas containing exhibits and needed to operate with either doors open or closed.

Two primary return points were selected, one the kitchen extract at the east end of the building and the other near to the main entrance doors at the west end. The latter causes an air flow in the opposite direction to incoming visitors and produces a variable temperature condition in the main entrance hall, where there are no exhibits. In summer the temperature at the entrance is higher than that for exhibits and this reduces the risk of thermal shock which might be caused by people entering from a very warm external condition.

The quantity of fresh air needed to reduce odours to an acceptable level is proportional to the number of people present and the design permits the quantity of fresh air to be manually selected by the remote operation of automatic dampers on the air plant connections.

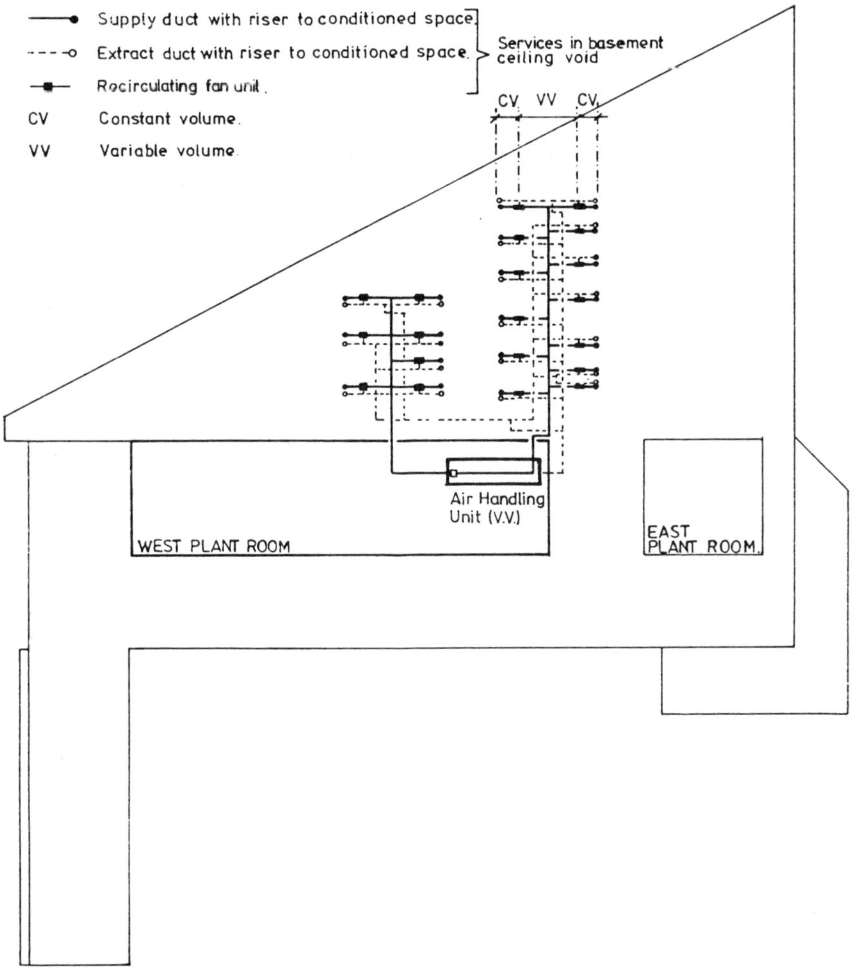

Figure 8.2 *Air system location*

The need for directional air flow with variable fresh air produced four main conditions. These are for day and night operation each with minimum and maximum fresh air. Figure 8.3 gives the air balance for the day condition with minimum fresh air. This Figure also shows the function of the transfer air duct which has been incorporated to distribute surplus fresh air evenly around the Gallery and to obtain the required directional air flows without the need to discharge surplus air quantities to outside.

Energy considerations

The factors that determine the energy consumption in a gallery are many. They include the number of people present, which can vary from almost no-one, certainly no-one when the building is closed; the extent and power of the display lighting, which again is off when the building is closed, but also changes in type and power as the exhibits are changed around and/or brought in and out of storage; the efficacy of the solar blind system in two ways — one the way in which the staff prefer to have the sunlight or not and the other the extent to which the blinds are operational and, of course the variations of the external weather conditions.

Because the Burrell has to have temperature and humidity control 24 hours per day 365 days per annum in almost every space of the building, whether people are present or not, the potential energy consumption looked formidable. It was decided that full advantage should be taken of the available humidity and temperature tolerances to reduce the energy consumption to the least feasible while keeping the conditions within tolerance. This is achieved by switching the fans off when the Gallery closes and all the display lights are switched off. The fans remain off until the temperature rises, or falls to the full extent of the permissible tolerance at the sensor. When the full extent of the permissible tolerance is reached the fans restart and the system modifies the internal temperature to the other extreme of the permissible range. Then the fans switch off again and so on until the Gallery reopens.

Common sense said that there must be many times when the external conditions were close to the required internal conditions and calculations using the then relatively new BRE theory of admittance suggested it could lead to the fans only running for 50% of the time from which massive energy savings would be achieved in the electricity used to run the fans alone.

Energy audit prediction

In order to evaluate all of these variables a total energy audit calculation for each month of the year was prepared using the Electricity Council Building Energy Estimation (BEEP) program. Because it was not possible to validate the BEEP program results a manual analysis was also prepared. This was based on maximum heat loads produced using an in house computer with a program specifically written by our Practice. The significant factors that emerged from this study were that the essential costs of energy for fans, lighting and mechanical refrigeration exceeded the cost of space heating. In the detailed design the energy saving factors that were considered were double and triple glazing, high thermal insulation of the roof and solid walls, the limitation of fresh air quantities compatible with the occupancy in the exhibition areas and the use of free cooling in other areas as much as possible.

The estimated energy consumption, for the average or mean condition, which was selected as the main design basis, summarises:

Heating and hot water	1300 MWh	24%
Cooling	600 MWh	11%
Fans and pumps	2200 MWh	41%
Lighting	1100 MWh	20%
Catering and Miscellaneous	200 MWh	4%
	5400 MWh	100%

Energy consumption

It will be noted that the main energy consumption is by the fans and pumps and this summary assumes it is possible to switch the fans off for some 25% of the time. It was considered feasible that the actual off period could be up to 50% of the time because of the heat storage effects of the structure, the majority of the surfaces being thermally hard. That is there is a relatively high admittance. This enabled conditions to be maintained for long periods, without the system in operation, particularly when the heat gains and losses were low.

Since the actual energy consumption is running about 20% less than our predictions of the average energy consumption with this feature we believe it has been a success and that the fans, in fact, run for less that 50% of the time.

It has also been a success in maintaining the humidity at an acceptable tolerance. For one plant recently tested it was 57.5 ± 5.7%. This compares to a design requirement and setting between 55 and 60% with a tolerance of ± 5% RH. Since the control is within the set range and the tolerance excess of 0.7% is not measurable by instruments in common use, and certainly is of little significance, all achieved with the air plant switched off for at least 50% of the time, this is a gratifying result.

A temperature recording taken with low heat gains shows the plant ON period is indeed very small and yet the tolerance of plus or minus 1 degree centigrade is maintained.

Energy source, distribution and recovery

It was observed that, even with absorption refrigeration, 65% of the energy consumption needed to be electric in any event and detailed analysis led to the decision for an all electric energy supply. For heating, electricity is supplied to six (three standby) 100kW immersion heaters in three 14,000 litre thermal storage vessels and to two 155kW electrical input (615kW cooling output) refrigeration machines to operate a full heat recovery system.

Figure 8.4 is the energy distribution summary.

Figure 8.5 is the basic heat recovery circuit employed.

Table 8.1 gives the month by month comparison of estimated and actual loads and consumptions, given in MWh rather than GJ in order to give direct comparison to the electricity accounts.

Integrating the hardware

Good gallery practice demands air supply at high level with air removal at low level. The reason for this is that the clean incoming air should be the main influence causing air movement around exhibits and dust collected by the air is most likely to be removed from the space by removal at low level. Certainly the reverse would cause dirt carried in by people and created by the wear of the floor surface being carried over the exhibits and must be avoided. Construction from the point of view of the accommodation of air ducts is a reinforced concrete basement with concrete ground floor with pairs of reinforced concrete beams downstanding into the basement. Rising from this structure there are reinforced concrete columns supporting primary beams of laminated timber. Secondary beams cross these and support the flat roof. At the bottom between the beams there is a stone curbing on top of which plastered brick panel walls form a double wall construction and give a rectangular return air entry point. The air supply ducts rise between the two primary beams, run horizontally and connect to stub ducts and supply air grilles located between the secondary beams.

It is as close to the ideal form of gallery air distribution pattern as I

MONTH		JAN	FEB	MAR	APR	MAY	JUN	JUL	AUG	SEP	OCT	NOV	DEC	TOTALS
MAXIMUM DEMAND														
Anticipated														
Max	kW	1584	1544	1565	1447	1438	1448	1349	1364	1332	1421	1471	1465	
Mean	kW	1351	1305	1306	1170	1131	1118	1010	1028	1014	1123	1213	1226	
Min	kW	1113	1064	1041	890	821	784	668	688	692	820	950	983	
Actual														
1982	kW					360	670	690	899	940	1000	1059	1130	
1983	kW	1099	1094	989	1000	1000	959	860	740	800	1000	1000	920	
1984	kW	1100	1000	1120	940	959	939	989	1000	910	820	959	940	
1985	kW	959	1000	900	859	889	940	959	955	821	799	800	839	
1986	kW	819	799	820	719	679	720	800	799	820	979	979	799	
CONSUMPTION														
Anticiptated														
Max	MWh	579	574	583	567	576	585	571	571	560	570	566	561	6863
Mean	MWh	504	489	489	447	434	428	396	401	397	432	461	466	5344
Min	MWh	429	404	392	325	293	272	221	230	234	293	356	372	3821
Actual														
1982	MWh					241	177	125	164	282	509	415	557	2470
1983	MWh	526	459	440	256	406	477	331	279	358	375	375	417	4699
1984	MWh	611	341	541	431	338	411	365	346	393	360	390	487	5014
1985	MWh	529	404	460	319	336	395	240	290	292	333	244	385	4227
1986	MWh	336	359	331	258	230	301	271	241	292	320	246	291	3476

Table 8.1 *Burrell collection building: Electricity consumption analysis*

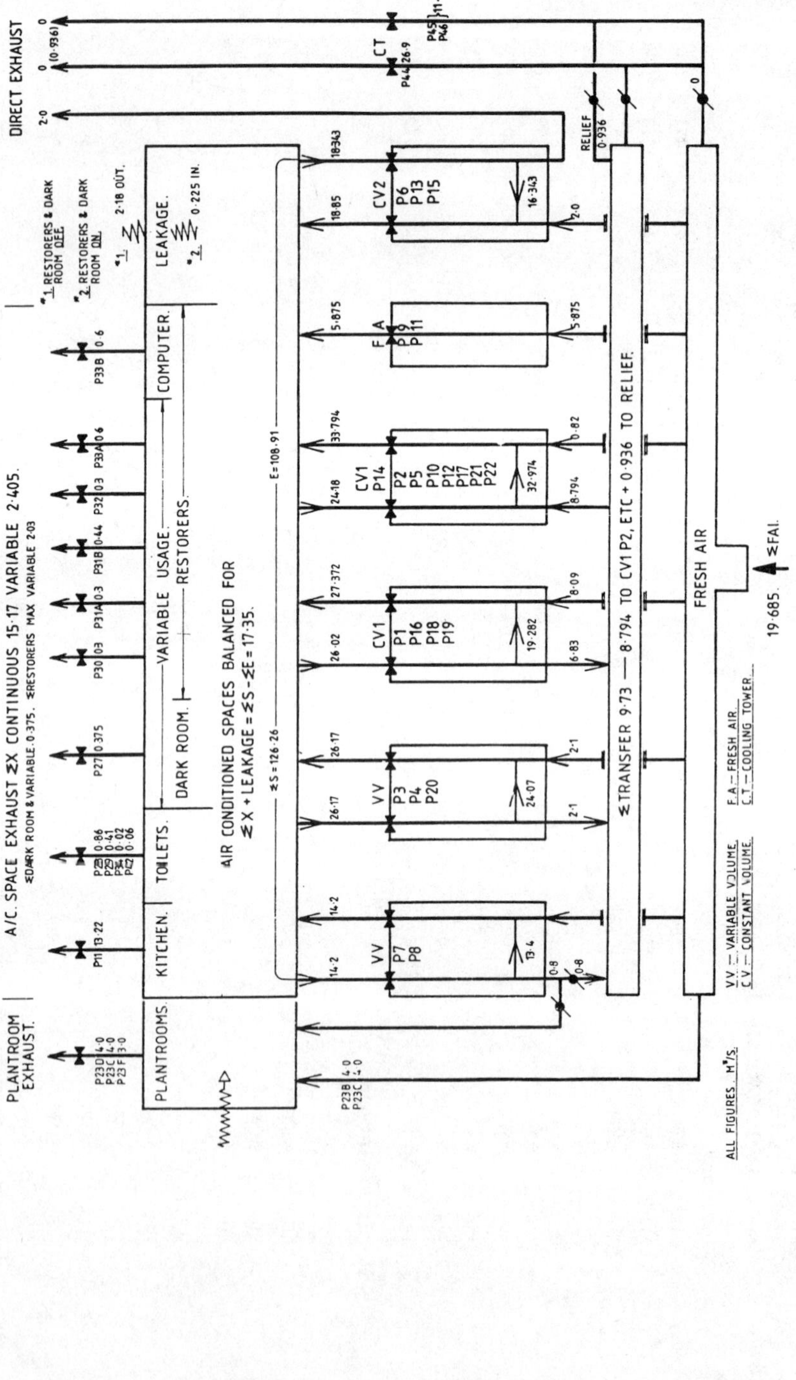

Figure 8.3 *Air flow balance*

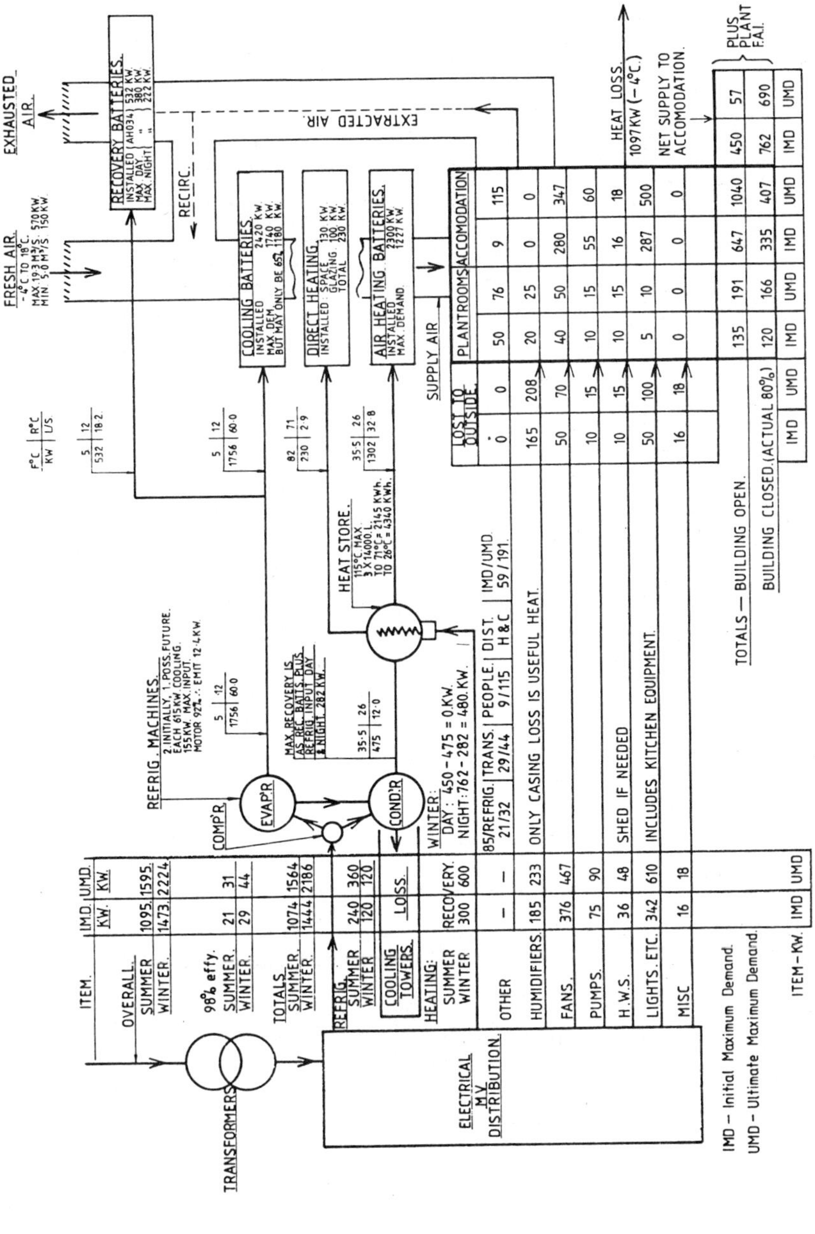

Figure 8.4 *Energy distribution*

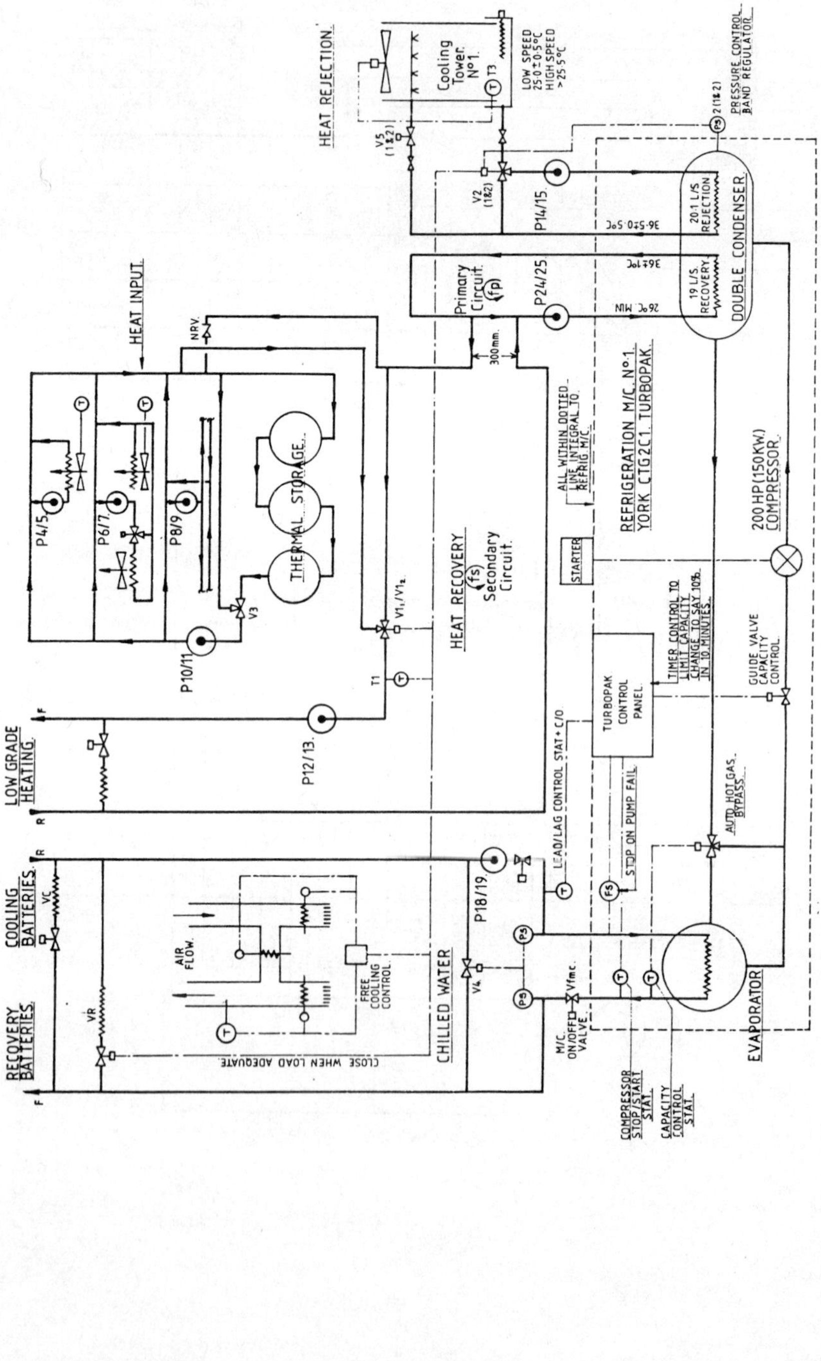

Figure 8.5 *Heat recovery*

believe it is possible to get and yet keeps strictly within the architectural concept.

The main supply and extract air ducts run horizontally in a false ceiling at basement level and connect between the central core basement plant room to ducts in the double walls.

The essential thrust to achieve the basic simplicity of the original conception came from Barry and the achievement of actually getting it to work, was, I like to think, because both structural and services engineers fully appreciated the architectural objectives and arranged their own bits and pieces so that they would mutually pin together and not intrude more than essential.

Solar protection and electric illumination

Illumination is controlled by internal and external blinds to glazing and by luminaires under full dimmer control for artificial lighting for display and general lighting in the absence of adequate natural lighting.

Building services monitoring computer

The engineering services are monitored, and to some extent controlled by a computerised building management system located in the central security control room. This room also contains security and fire alarm panels, central sound broadcast station and closed circuit television monitors.

The role of the Quantity Surveyor

In addition to the Architect and the two engineers, nearly all of our detail design meetings were also attended by the Quantity Surveyor. This is unusual I think, but felt by the Burrell team to be essential. Victor Bugg clearly had his own role in being responsible for producing the bills of quantities and other contract documentation which would form the basis of the contract between the Contractor and Employer for the construction of the building, but was also invaluable in warning where particular design solutions would lead to excessive cost, both inherently in the materials involved and, less obviously, in the effect it would have on the constructional sequence, or at least the necessity to point out to the Contractor the sequential consequences of the design.

One example of this was the method of installing the air ducts between the double skinned wall. The vertical air ducts were easy enough because one simply built one of the brick leaves first, then installed the air ducts, tested these for air tightness, applied the thermal insulation and vapour barrier and then sealed the lot in by building the second brick leaf.

The difficulty arose at the top where the flat roof is supported on secondary beams passing at right angles across the double skinned walls. These beams had to be installed before the roof coulf go on and I was not happy to see air ducts installed until the building was reasonably watertight. That meant the secondary beams needed supporting before the installation of air ducts and this is by primary beams running along the tops of the two leaves of the double walls. Because inset show cases run up to the underside of the primary beams the horizontal air ducts have to run between the pair of primary beams and had to be installed, fully insulated and tested as insulation subsequent to installation would not have been possible.

Clearly this is a matter no contractual team could have been expected to have fully appreciated at tender stage unless drawn to their attention.

The total design: *Barry Gasson*

Collection and site

Sir William Burrell (1861–1958) was a Glaswegian shipowner who presented to his native city the collection of works of art that he had acquired over a very long lifetime. It is now known as the Burrell Collection. In return the City agreed to provide a new building so that it might be both enjoyed and preserved for time to come.

Burrell collected from Europe, the Middle and the Far East, and the Collection comprises stone windows and doorways, carved screens, a timber ceiling, furniture, timber and stone sculptures, tapestries and carpets, needlework and lace, stained glass, paintings and drawings, and many objects of stone, timber, metal, porcelain and glass. It has breadth of time, of scale and of subject, yet it has integrity. It also contains pieces out of which one can make a building, and the three rooms from his home at Hutton Castle that he requested to be reproduced.

The site is a field surrounded by trees and woods in Pollok Park, the country estate of Pollok House, some three miles south west of Glasgow.

Figure 8.6 *The North Galleries*

Design objectives

The building at this point in time we felt had to bridge a cultural gap probably wider than ever before and this directed those initial aims.

We considered that within the building each object should have a feeling

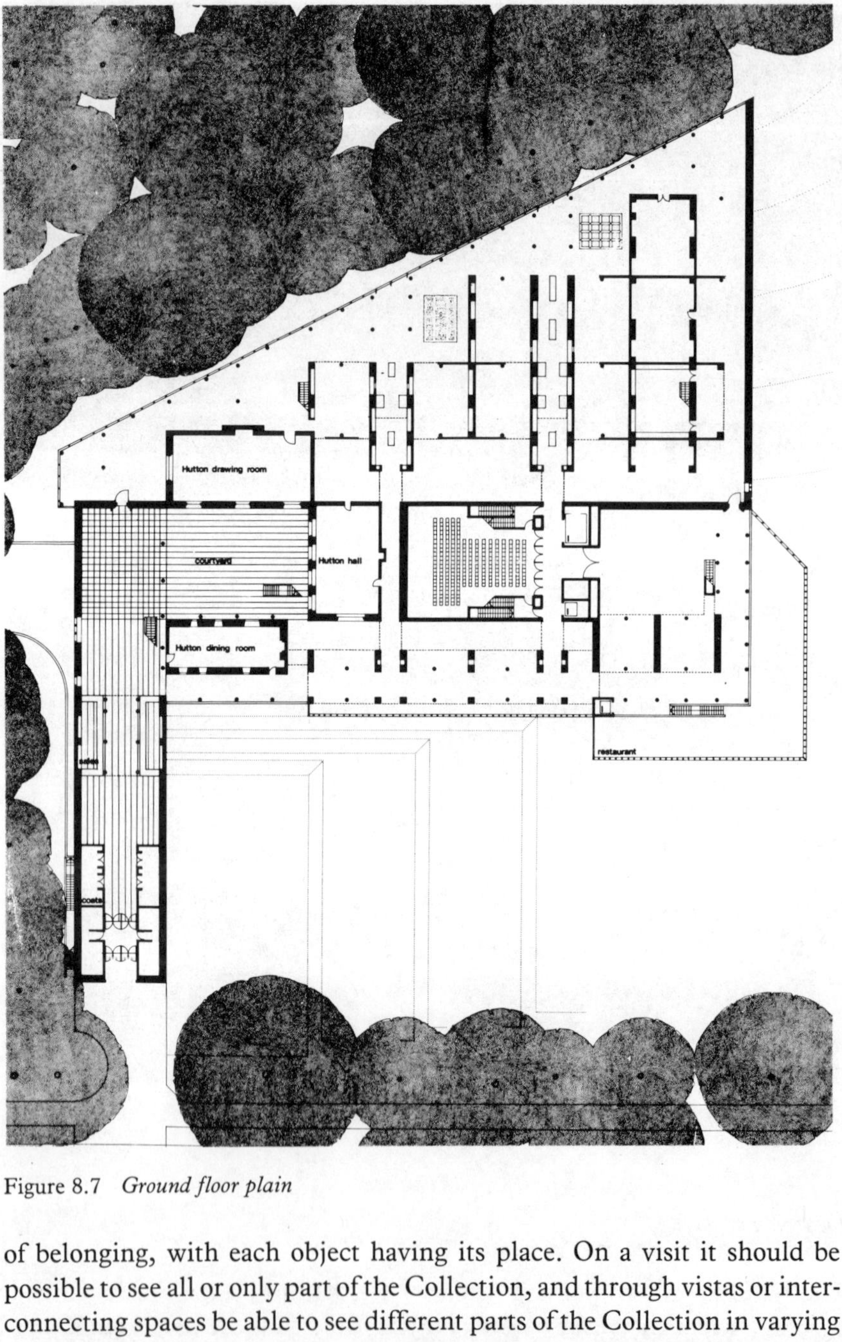

Figure 8.7 *Ground floor plain*

of belonging, with each object having its place. On a visit it should be possible to see all or only part of the Collection, and through vistas or interconnecting spaces be able to see different parts of the Collection in varying relationships with each other, and comprehend the integrity of the whole.

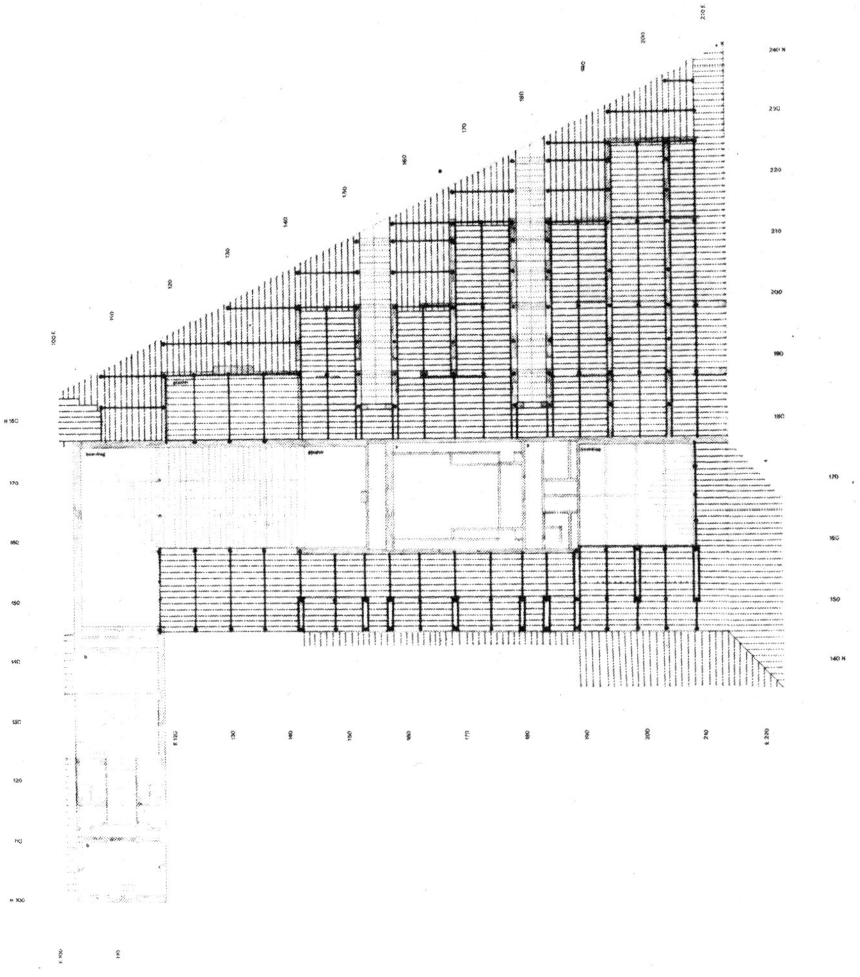

Figure 8.8 *Roof plan of timbers*

We also considered it fundamental that the external world should penetrate the building and bring into direct relation with the Collection the changing seasons and the changing day. This would help to make the objects both a delight to enjoy and to suggest the essence of their being.

Implications

These considerations generated a building and a display that was primarily on one level, that of the ground and the surrounding park and woods, and one with a semi-open plan form.

The variety of the objects in the Collection generated the needs for different qualities of space and of light; there were objects that would enjoy the changing pattern of natural light and those that had to be protected from it artificially lit.

The materials that were used within the building were those with timeless and traditional associations; floors are of stone, with areas of timber or carpet, the walls are plastered brickwork and over the whole spans a laminated timber ceiling.

Services

Into this building fabrics are woven the complex systems of services that control the light and the air, and that provide protection.

There are the issues of controlling the quality of air in a building that has an open plan form, has a variety of different kinds of space characteristics, is of mass materials, has a glazed edge, a deep plan form, and above all has a collection of objects that range from stone and metal to fabrics and painted timber panels.

Yet there are also the formal issues of how does one integrate within a building that displays objects created over the last three and a half thousand years those systems that are a manifestation of todays needs, very much out of time and place, yet have a contribution to make in their own right in the making of total building that not only aims to enhance the enjoyment of the Collection but is concerned with art in its own right.

What we have achieved is best shown by the pictures and figures.

Structure and services: *Victor Dowsett*

Superstructure

In heavily serviced buildings, such as the Burrell, the structural concept has to be developed in parallel with the services, is greatly influenced by the service requirements, and must be under the firm control of the architect.

The north to south distribution of services from the plant rooms occurs in the basement ceiling void, with the main ducts located between the floor beams which run north to south only so as to leave a clear path for the ducts.

On alternate grids, a system of double beams are provided at ground floor and roof level, thus creating vertical circulation routes between the beams for the services. The floors are in reinforced concrete. The roof beams are in laminated timber, and support a timber plank deck, into which cable trays for electrical services have been integrated.

The columns, and the grid of main and secondary roof beams have been compared to the trunks and canopy of the adjoining woodland, with the energy rising through the services being akin to the sap in the trees!

The basic minimum structure for the roof would have been simple single main and secondary beams supporting the timber deck, with the size of the beams adjusted to their varied spans. The give and take between structure, services and architecture has resulted in the following adjustments to the basic minimum:

- Paired beams on alternate grids.
- Standardised beam sizes irrespective of span.
- Design of connections between timber components to allow passage of services.
- Integration of cable trays into timber decking.

Mention has already been made of service problems related to the large areas of glazing in roofs and walls. An unobstructed flow of air across the inner face of the glass was required to control condensation, while on the other hand, bracing was needed to some of the slender timber sections which support the glazing. Some form of slender metal section was therefore proposed for the bracing, but we were concerned that high temperatures could occur in the metal just beneath the glass, leading to excessive thermal movement in the longer bracing members, if normal structural steels were used. The final solution comprised 25 & 32mm dia INVAR steel rods which have a zero coefficient of thermal expansion and which offered little obstruction to the air flow.

We had been concerned at design stage, that timber members directly beneath the glass, particularly on the south face, might be damaged due to excessive heat and drying out caused by solar radiation before the external blind system became operative, and the internal temperature and humidity levels were fully under control.

The specification called for shading of the glass until the blinds were operative and it had been envisaged that some form of wash or film could be applied to the glass as a temporary measure. This proved impractical due to the risk of damage to the double glazed units from differential temperature conditions in the glass skins.

It was therefore decided to monitor conditions in the timer at two typical positions, so as to assess the likely magnitude of the problem.

Thermocouples were installed just beneath the surface of sample blocks, fixed to the timber on the most exposed faces. Measurements of outside and inside temperature and humidity together with the surface temperature of the timber and it's moisture contents, were taken over a month period in June–July 1981. Weather conditions at the time of each reading were also recorded.

The highest surface temperature reached was 40°C at 2.00p.m. on a sunny afternoon when the air temperature was 22°C and the relative humidity was 57%. Throughout the test period the moisture content of the timber remained static at around 13%, so that it would appear that the Glasgow weather did not provide sufficiently long periods of sunshine to cause damage to the timber.

Basement Construction

The aim of the Structural Engineer is to achieve a very simple profile to the external basement envelope for the following main reasons:

- To provide the best conditions for the construction of 'water excluding' concrete.
- To facilitate the practical design and placing of external waterproof membrances where these are required.
- To speed construction.

If our aim is to be achieved it is important for service requirements that will influence the basement profile to be available at the concept stage of the design process. The give and take dialogue between the design team can then begin, which will hopefully lead to the most efficient and cost-effective total scheme.

The late introduction of additional service ducts which disrupt the agreed profile may make nonsense of the earlier decisions with respect to structure. It must also be said that late changes to the structure may have similar effect on the services.

It is also important to establish at the concept stage, the degree of protection that will be required against penetration through the structure of ground water, or water vapour. There are three alternative degrees of protection (4):

- Utility Grade
- Habitable Grade
- Special Grade

The decision may well influence, or be influenced by, the ability of the Air Conditioning to control moisture entering the building through the structure either in liquid or vapour form.

The main plant rooms for the Burrell are situated in the basement and, together with the primary air supply ducts, occupy a large proportion of the basement volume.

The locating of the refrigeration plant, and the primary north to south air supply ducts at the Burrell are examples of the give & take process previously described.

The basic requirements were:

- A location for the refrigeration plant which would prevent noise break-out to the public areas.
- A primary north to south air supply route from the intake zone along the south face to the east to west duct running along the north side of the main plant rooms.
- The structural engineer's aims as previously described.

Among the alternative locations considered for the refrigeration plant were:

- A site remote from the main building.
- A position within the main plant rooms.
- A position beneath the main plant rooms.

The remote site was economic structurally, but was not favoured due to the length of pipework between the plant and the main building.

The position within the main plant caused no structural problems, but noise control would have been difficult and plant room space at basement level was at a premium.

The position beneath the main plant was contrary to our aim for simple basement profiles, and involved the most expensive structure. Nevertheless, from the over all point of view, it proved to be the best scheme for the following reasons:

- Short pipe runs to the cooling towers which had to be located adjacent to the lifts.

- Noise break-out controlled by remoteness of position relative to the public areas.
- The primary north to south air supply route could be located alongside the 'refrigeration plant room, thus at least in part, simplifying the basement profile.

Conclusion

It may appear from the quoted examples of structure-service-integration at the Burrell, that it has been all give on the part of the structure relative to service requirements. On balance there is probably more give than take, but it must be remembered that the cost of the basic structure at the Burrell is some 16% of the total, and thus modest increases in structure can usually be accommodated without undue effect on the total budget, and are certainly necessary in order to achieve the total design concept.

The building that can now be seen in Pollok Park has not in essence changed from the competition entry. Credit for this must go to Barry Gasson, whose single-minded approach has held us to his vision throughout the project.

Managing the contract: *Victor Bugg*

The Contract

The City Architect at one of the earliest briefing meetings made a number of points.

The exhibits were irreplaceable, could be damaged in many ways, heat light, dust, damp, physical etc., and yet were to be integrated in their totality within the gallery before the building contract could be considered complete.

The building would be opened and in use by the public before the building contract could be completed.

Everything necessary for the completed operation, including furniture, fittings, showcases, typewriters, would be part of the contract.

The period for the setting-up of the collection would require occupation by the museum's staff, totally controlled conditions and many contractors workmen in all trades working in parallel.

The City at that time had their own form of building contract, and the normal practice was to make considerable use of nominated sub-contractors and suppliers.

This time said the City Architect, we want total control by the main contractor, with no nominations and we want to use the standard form of building contract for the first time . . . oh and, by the way, we shall use our own clerk of works.

So the conditions of contract were born.

JCT Standard Form of Building Contract 1963 (1976 revision) with quantities with some nominations on the basis of Employer Sub-Contractor agreement.

Everything to be measured including the mechanical and electrical.

Nomination would be made in respect of the controls, the control centre and the protective installations primarily because a design input from the specialists was essential.

The appendix to the form of Contract included the following: —

Defects liability period	12 months
Date for possession	1 April 1978
Date for completion	13 October 1981

The Contract Documents

The contract documents included inter alià, the following clauses:

Description of works

The Contract includes for the construction and installation of the fittings and furnishings necessary to house the exhibits forming the Burrell Collection. The Contract also includes for the incorporation into the fabric of the building various stone items and wood ceilings and panelling. A number of rooms are to be near reproductions of rooms originally existing in Hutton Castle.

It will be necessary for the mechanical and electrical installations to be substantially operating three months prior to the date of practical completion.

Comprehensive electrical and mechanical engineering systems will be provided throughout the building to control natural and artificial lighting and air temperature, humidity, cleanness and movement; and to protect the Collection against damage or loss by fire, theft or vandalism. Fire and security detection and automatic fire fighting systems will be provided.

The basement plant rooms will contain high voltage sub-station and switch room, control centre, thermal storage, and water chilling plants, air supply plants, bulk storage, carbon dioxide fire system, standby generator etc. Local air temperature will be maintained by air terminal units integrated at high levels within the basement, and other points in the building, within false ceilings and in wall spaces. Lighting levels will be adjusted from local panels within the exhibition galleries and elsewhere.

The building fabric has been arranged to permit the careful integration of the mechanical and electrical engineering systems in space provided within the walls, ceilings and floors of the various zones to permit the passage and accommodation of the various duct and conduit systems and their terminals.

Shop drawings

The term shop drawings as used herein includes fabrication, erection and layout drawings; manufacturers' standard drawings; schedules; descriptive literature, catalogues and brochures; performance and test data; wiring and control diagrams; all other drawings and descriptive data pertaining to materials, equipment, piping, duct and conduit systems, and methods of construction as may be required to show that the materials, equipment or systems and the position thereof conforms to the Contract requirements, or as may be required in any specification or bills of quantities incorporated in any sub-contract under Clause 27 or any Contract for the supply of materials under Clause 28 of this Condition. As used herein, the term manufactured applies to standard units usually mass produced and fabricated means items specifically assembled or made out of selected materials to meet individual design requirements. Shop drawings shall establish the actual detail of all manufactured or fabricated items.

It shall be the responsibility of the Contractor to verify all dimensions from shop drawings and site conditions and to co-ordinate the work of any sub-contractor or materials supplier with the work of any other sub-contractor or materials supplier whose work is related thereto.

Practical completion and defects liability

All engineering systems are to be commissioned and in full operation three months prior to the date of Practical Completion only except such items as cannot be completed until the Museum display and other objects are put into their final positions.

Co-ordination of the whole of the works

The Contractor shall be responsible for the co-ordination and proper execution and completion of the whole of the works and the protection of the site.

The Contractor's attention is particularly drawn to the nature of this Contract which will require site management which is fully experienced in the co-ordination and direction of his sub-contractors.

The Contractor will be provided with detailed co-ordinated engineering services installation and setting out information showing the critical relationships of engineering and building elements for major duct routes throughout the building.

The drawings contain sufficient information for the specialist sub-contractors, individually and separately to produce such additional control wiring diagrams, shop manufacturing drawings and to proceed to installation with no references to other sub-contractors except insofar as sequence of installation is concerned. A competent and experienced co-ordinating engineer shall be allocated to this project for such time as may be necessary to fully dispense the Contractor's responsibilities set out in these preliminary and general clauses.

On the site

The building contract did not run exactly as programmed, and the original dates became extended.

Original completion (13 October 1981) date became:

a) Partial possession	July 1982
b) More possession	August 1982
c) Practical completion date (heavily qualified)	March 1983
d) Royal opening	October 1983

Observations

A few observations during the progress of the work included:

A hostility by the sub-contractors to a domestic sub-contractor role.

The contractors very early decision to virtually revise his planned programme of works to take advantage of what seemed to be an exceptionally dry summer. The effects were far reaching because the summer was not dry and the presence of water caused much disruption.

A continuous dialogue between contractor and sub-contractors as to who was co-ordinating the commissioning of the air-conditioning services and despite the contract clauses the difficulties of securing total commissioning.

The incurring of costs not always recoverable substantially beyond anything envisaged by the contractor in the commissioning process.

The architects insistence on surfaces free from projections etc. imposed at times a difficult discipline on the installers.

The conflict between dust conscious sub-contractors (electronic equipment) and plasterers, etc. working at the same time.

The dearth of capable co-ordinating engineers who read drawings before commencing installation.

The anomaly, during a high inflation period, of the application of the NEDO formula one equipment being brought too early on to the site and then applying the formula at the time of fixing.

Despite the difficulties, the obvious pride and enthusiasm of all the craftsmen, technicians and others engaged on the project.

References

1 Michael Brawne *et al.* The Burrell Collection. *The Architects Journal* October 1983.
2 Stephen Ashley. Making the Burrell Work. *The Journal of the Chartered Institution of Building Services* December 1983.
3 Barry Gasson Architects. Burrell Museum, Glasgow. *Architectural Review* February 1984.
4 Circa Guide 5 — *Guide to the design of waterproof basements.*

9 Refurbishment of Unilever House

Introduction: *J.H. Medworth*

Unilever was founded in 1930 when two businesses which were already established in many countries, Lever Brothers in the United Kingdom and the Margarine Unie in the Netherlands, came together.

Over 300,000 people are employed in more than 500 subsidiaries in some 75 countries. The operating companies supply a wide variety of goods for everyday use.

Unilever is a major producer of consumer goods: mainly food and drinks, detergents and toilet preparations.

Despite its size and international significance, the name Unilever is relatively unknown among the public at large. Millions of people buy Blue Band margarine, Cornetto ice cream, Lipton Tea, Surf washing powder, Sunlight and Lux soaps and Signal toothpaste and a multitude of other well known products without even seeing the name Unilever.

Since the time of the original merger between the Margarine Unie and Lever Brothers, Unilever has had two parent Companies, now called Unilever N.V. Rotterdam, and Unilever PLC, London.

History

Unilever House, Unilever's London Head Office is a prominent landmark (on the Thames Embankment) overlooking Blackfriars Bridge, has been the headquarters of Unilever for half a century. It is easily recognised by its curved Ionic collonades and Portland stone facade.

The building was constructed in the early 1930's at the time of the merger between Lever Brothers and the Dutch counterpart Unie Company on the site previously occupied by the De Keyser Royal Hotel. It was designed by Unilevers' Chief Architect, Lomax Simpson and implemented by Sir John Burnett Tate & Lorne.

The building opened in 1932 and for its day it was probably the last word in modern office design. The original interior design contained elements of Art Deco, particularly in the old main entrance and this Art Deco style formed the springboard for the renovation strategy.

Inevitably, the building, offices and particularly its services became outdated and in the late 1960's Unilevers property experts were asked to consider the companys long term future accommodation needs and for 5 years studies into the various options open to the company were carried out.

Survey of staff preferences were undertaken. The special study produced three main options:

- New purpose built headquarters out of London
- Redevelop the Blackfriars Site
- To Refurbish Unilever House

It was decided that the company's best course was to stay in London.

Expansion of the headquarters over the years had necessitated the acquisition of a number of nearby Blackfriars properties. About 3500 Unilever staff worked in the various buildings at one time however a policy of gradual decentralisation was to reduce this number to a core staff of about 1500 by the late 1980's.

In order to achieve this within the bounds of a single head office building renovation had to achieve greater use of space available — almost half the gross area was taken up with entrance halls, corridors toilets and other non occupiable areas.

The main building provided a usable space of 18,000m^2.

Detailed plans were put into action to renovate Unilever House retaining the original facade — which by 1977 had been listed as being of historic interest, and demolishing neighbouring older buildings and replacing them with a new eight floor North Wing giving a further 6,500m^2.

The renovation of Unilever House, was an ambitious project particularly as it was to be achieved with the building still occupied.

Contractor selection

Over 40 contractors were screened for their suitability in carrying out what at the time was considered to be the largest refurbishment of an occupied building in the whole of Europe. 10 were short-listed and these were sent enquiry forms requesting information relating to:

- Previous Renovation Experience
- Financial Standing
- Contract Commitments
- Available Resources
- Record for completion of Programme
- In-house trade expertise.

From these ten, five contractors were invited to tender, contractors were requested to visit Unilever House where they were briefed as fully as possible concerning the project with particular emphasis on site working, storage and access.

The two lowest tenders were invited to present their proposals for organising and surveying the project together with the personnel that would be involved on site. As a result Wates Special Works Ltd were appointed.

There were many nominated sub contractors appointed during the course of the contract and this insured that the expertise and craftsmanship required on a project of this size was achieved.

Project management

A board director was put in overall charge of the operation assisted by one of the Unilever senior management as Renovation Controller. As well as Unilever's own design team, a consultant architect, Theo Crosby of Pentagram was engaged to advise on various aspects of the work.

There were two key objectives. One was to increase the amount of first class office space inside Unilever House which involved installing air conditioning and making a number of suitable changes to the outside of the building: adding front windows at eighth floor level to create a new suite of offices for the companys top executives, lowering the parapet obstructing the seventh floor offices, cutting windows at the rear of the building and removing cooling towers from the eighth floor. Eventually to draw attention away from the new windows 14 twice-life size satues were positioned on the seventh floor parapet above the pillars of the colonnade.

The second objective was to devise and apply a consistent strategy of visual design inside Unilever House. The theme related closely to the Art

Deco styling which was the vogue when Unilever House was built at the end of the 1920's.

Tackling the £37 million programme called for an enormous amount of Management planning and this was handled by forming a management team under the Renovation Controller.

For Unilever House a design team was formed led by Unilevers staff and helped by outside consultants Buckle and Partners and Gardiner Theobald. This team being responsible for project management architectural and structural engineering design, mechanical and electrical services design, quantity surveying and site supervision.

Much of Buckles work was carried out before detailed design was started. Their first appointment was to carry out a detailed feasibility study on how to bring the services up to scratch in the existing building and how to air condition the new wing.

Design concept

The North Wing consisted of a dual duct air conditioning plant, with a specially designed integrated ceiling, and hot and chilled water supplied from the main building.

The refurbishment of Unilever House provided new problems. The renovation of mechanical services whilst the user was in occupation presented major problems of planning, noise control, and maintenance of services in occupied areas.

Detailed discussions took place between client and consultant on areas to be vacated at any one time, type of air conditioning system to be installed particularly bearing in mind the need to minimise disruption and provide localised control economically. The system had to be capable of progressive extension as areas were renovated.

The plant eventually decided upon was a perimeter four-pipe system with fan coil units and ducted primary ventilation supply and extract air for office areas. Individual low velocity supply and extract air handling systems were supplied for other specialised areas such as conference suites, kitchen and catering areas, dining rooms, club, travel and communication areas.

In addition the provision of a fully air conditioned building required central refrigeration plant to be installed to meet the load. Spare boiler capacity was available from existing steam boiler plant for driving the two new absorption chillers. There were also two turbo chillers which provided flexibility of main energy source during limited electrical power availability.

Trying to locate this central plant within the confines of an existing building with maintained services where free space was at a premium proved a major problem.

The majority of the main plant was finally located in the basement with the remainder located at roof level following the resiting of the main water tanks.

Considerable investigation then took place to establish how the distribution of pipework and ductwork could be arranged from principal plant areas with minimum disruption to the building occupants.

To keep disruption on the occupied floors to a minimum the number of main vertical risers was limited to four, each serving a defined zone.

The need to fit plant into already existing areas resulted in an extremely economic use of space and at times rather more restricted than was desirable. The converse also being true when retained plant was finally removed.

Design and contract administration

Design Team meetings were held on a weekly basis to define and develop forward projects and to deal with problems that arose during construction.

Cost and Progress Meetings were held once a month to monitor budgets and progress to enable regular reports to be submitted to the Client Committee.

Site Meetings were held monthly with the main Contractor and nominated sub contractors.

Client Liaison Meetings were held weekly between Client, Contractors and Site Management to discuss and agree the organisation for future projects related to client access, maintaining means of escape, security, vacation of areas to allow continuity of work, disruption of engineering services and dealing with clients problems and complaints.

Principals Meetings were held informally once a month, between senior Unilever staff and two directors of the main contractors to discuss or anticipate any major problem.

Work planning

Throughout the 6 year programme a tight schedule was strictly observed — each sector involving 14 weeks work. A new zone was started every 4 weeks so that a complete floor was entirely renovated every 26 weeks.

To enable staff relocation to run smoothly advanced planning was essential and long before renovation work started detailed movement charts were devised to minimise disruption to both staff and contractors needing clear access to work areas whilst keeping the head office fully operational.

The problem of relocation was greatly eased by the completion of North Wing in 1981 which provided buffer space for staff from Unilever House.

The moving of staff into temporary offices was carried out in the main over the weekend in order to avoid disruption to normal working patterns. Departments were moved on block in order to maintain normal communication links as much as possible.

In order to organise the multiplicity of moves 4 in-house project teams were set up consisting of various disciplines to cover building, transportation and engineering. Their role was to liaise fully with the building occupants on all his needs and to effect a smooth transfer to both his temporary and then permanent accommodation. This on occasions meant two or more moves because of the complex nature of the programming. Overall costs of these moves were contained by encouraging departments to move into their allocated temporary accommodation with as little alteration to layout of partitions as possible. Even with these constraints costs for relocation of staff were in the region of £1.0m.

The programme of moves for final location of staff was maintained by ensuring that office layouts were frozen some 6 weeks before occupation not allowing any last minute changes of mind.

Site security and safety

Whilst the contractor was responsible for security of his own site the overall responsibility lay with Unilever. All contractors were issued with identity cards and building entry was strictly controlled.

Mobility of the contractor on site was controlled by partitioning off the work zones where possible.

Traffic up and down the building was restricted to a designated staircase of internal lift and the contractor whenever possible was encouraged to provide his own external passenger and/or goods hoist.

Control was not easy especially when it came to preventing contractors using vending machines, toilets and general lifts despite our insistence for him to supply his own amenities.

Linked with security was the question of safety on site. In view of the type of works being carried out in a fully occupied and operational building it was surprising how few accidents or major catastrophies occurred.

One can only put this down to the vigilance shown by all parties and in particular the contractors site safety manager, especially appointed, and our own security staff who carried out regular tours of inspection during and outside normal working hours. The implementation of a *Hot working permit* ensured that hazardous operations were monitored carefully.

Communications

Our normal fault report centre acted as the main communication links for both client and contractor. Key staff were issued with bleeps which enabled instant communication. Quick reaction to complaints was essential to maintain a good level of staff moral.

Lessons learnt

The refurbishment programme was a long and protracted project. At the beginning every situation seemed to be an insurmountable problem, but before long those involved with staff complaints or contractors immediate needs were able to put things into perspective.

To say it was a testing time for all those involved with the project is possibly understating the situation. This also went for those who were subjected to the traumas of refurbishments in an indirect way i.e. the office occupants. Staff tolerance to their ever changing environment grew as the contract went on. I can remember when the slightest amount of dirt and noise would bring building work to a standstill with threatened walkouts by staff. They soon learnt to accept the constant presence of the contractor and his strange ways. Some were even on first name terms by the end of the contract. The extent of the renovation work related to engineering services severely disrupted existing plant operations, however, at no time did the building suffer a major loss of service.

The real secret to the success of the project was:

- Good liaison at all levels
- Maintaining a strict programme
- A rigid budget control.

It taught me personally:

- That you can't afford to be an idealist
- That specifications need to be read carefully to make certain you pay for what you get and not get what you pay for.
- That refurbishment does not necessarily reduce running costs what it does is to contain cost within specified limits.

Financial implications

A cost comparison made between the new building and the refurbished section confirms that the theoretical calculations made which suggested that we should refurbish were born out in practice. Even without the site acquisition costs the North Wing which was a spec building of 6,500m^2 cost £1,916 per m^2 to build where as the refurbishment cost about £506 per m^2. With the anticipated sale of the buildings located in Blackfriars realising £28M on paper the six year operation will have cost Unilever £9M net. This in property terms would buy approximately 15 months of rented office accommodation of the proportion held within the city area. Refurbishment obviously made good economic sense.

Practicalities of air conditioning occupied buildings: *R.C. Maynard*

The refurbishment of Unilever House was both interesting and a very demanding project to the Mechanical Contractor for not only was the Client in full occupation during the process but also the Client's full Design Team and so the installation works were subject to constant inspection by all Parties.

To allow the installation of the air conditioning within this occupied building the works were sub-divided into two distinct and separate phases.

The first being the provision of the main air handling plant and the distribution heating and chilled water pipework alongside the distribution ductwork.

Having installed these services the next phase of the works could proceed and although the works on the next phase commenced on the top floors and gradually worked its way down the building, with flexibility of this type of design this was not essential.

Returning to the first of these works, the installation of the mechanical plant, the first essential items was the conversion and adaption of the existing building to provide areas for the new plant and the main service risers to the four section of the building.

To enable this to be achieved, the main cold water storage roof tank room was enlarged and converted into the air handling unit plantrooms and five new low profile sectional tanks erected on the existing roof and covered by a lightweight pitched roof.

These works on the roof were carried out in a phased manner to enable gradual changeover from the existing to the new domestic services concluding with a new plantroom area to provide the primary conditioned air to the eight office floors.

The distribution of services were installed down a combination of lightwells, converted lift shafts and redundant pipework riser shaft that were made available from the redesign of the building.

The remainder of the service areas for the building within this category, we include the Kitchen, Catering, Conference, Postal, Travel, Communications, Maintenance and Store, are generally located on the lower floors and are therefore ideally located to be fed by the various services from the plantrooms and boilerhouse in the lower ground floor basements, this also necessitated the modification of a large furniture store into plantroom area.

Having installed the plant and main distribution ducts and pipework, the next phase could proceed, which was the gradual replacement and installation of new services to the eight floors of the offices, which included new toilet facilities and provision of the air conditioning to the refurbished offices.

Having summarised the installation works, we turn to the practical problems in the two phases and the demands on the installation team.

The first phase included, as previously stated, the adaption and modification of certain areas to enable the plant to be installed. Being an occupied building these major building demolition works were carried out during early morning, lunch time and during the evening.

During these modification works the main ductwork and the heating and chilled pipework were installed and as it had no connection with the existing services, the project proceeded in a manner very similar to a conventional new building, brackets were installed and the various services erected with connections being left at the appropriate locations on each floor for future extension.

Once the roof and basement plantrooms had been formed or adapted, the supply air handling plant and remote extract fans could be delivered to site. The roof plant, as previously explained, was formed from an operational tank room, to allow this to occur a new tank area was formed and two new sectional GRP tanks erected to allow the first changeover to occur. These changeovers and adaptions of new and existing services were carried out over weekends and evenings when the building was unoccupied.

Numerous adaptions of services occurred throughout this project by a dedicated team of fitters, fitter welders, duct erectors, electricians, builders and plumbers, for once commenced the adaptions had always to conclude by 9.00 Monday morning with a fully functioning office block.

Fortunately, the majority of plant these days can easily be supplied in sections, and this was put to good use in the delivery and manhandling of plant into the roof plantroom.

The lower plantrooms are located below ground in a sub-basement level and to gain access the Client vacated an internal car park which provided both a home for the site management team, small stores for all disciplines and once opened up, the way into the plantrooms. This car park was the nucleus of all planning and access onto the site, together with the Client's security check for all the men on site.

The sub-basement plantroom also housed the Client's existing plant, this was removed and replaced by new air handling units and fans which were served from the new chiller station and existing boilerhouse.

To enable these units to be installed, motors and fans were removed from their respective subframes and lowered into the plantroom following the erection of lifting beams and support frames.

By this stage of the project the new chilled water facility and existing boilerhouse heating was available for extension into the new air handling plantrooms and, therefore, on a phased basis, the lower areas were gradually brought on to line in accordance with the building refurbishment. These areas ranged from full Catering and Dining facilities, Conference & Lecture Theatres, Premises Services Department, Car Park Ventilation and Chauffeurs Rest Areas.

During the first phase of the works an essential section of works took place titled the Demonstration Area. A half section of one of the zones was completely refurbished in accordance with the final design concepts. This enabled both the design and installation team a complete practice run and upon completion a complete review took place. To everyone's credit very few changes occurred from the design envisaged, but it gave all parties the ideal opportunity to have a learning area. Finally, of course, the Client's own maintenance staff visited the area and made their own judgement.

By this time the first phase of the work was drawing to a close and the actual refurbishment of the office areas could take place.

At this point of the contract a different type of work started to take place which required a different type of labour force and we therefore moved off from the site our heavy team, the welders and plant erectors, and substituted a second fix team of men who would be able to work in the building, in some cases alongside the Client's office personnel including Managers and Directors. This was absolutely essential to ensure the smooth running of the works and much closer inter-relationship with the Resident Architectural and Design Teams. Prior to this our work could, in general terms, be considered as a normal installation.

From the demonstration area we realised that to ensure a smooth running installation a programme of works was required to enable continuity of all the specialised operations and following detailed discussions with the building contractor this working arrangement was formalised with all parties.

The only problem remaining, however, was that the design of these areas was to be released in a gradual manner, this presented a number of problems. Firstly, the storage space of materials was minimal and although the basic materials had been declared, a system had to be developed so that upon release of each zone the various sections would site survey the areas. From this the working and manufacturing drawings could be prepared, showing ducting and pipework layouts. In parallel to this a complete take off of materials was scheduled and placed on order, all this took place some six weeks prior to the start date on site.

During these weeks a routine was developed which included full submission of detail layout drawings to the Client for approval and a well worn track was developed between the respective offices for this was the method employed over the next two years during which time 28 zones of this building were completely refurbished. The approximate value of each of these zones being £150,000.00. The site surveys were carried out in the occupied office areas and this necessitated the cutting of access panels through the false ceilings to establish whether the design duct and pipe routes could be maintained, and the removal of access panels to establish routes of existing pipe and duct routes, all of which were replaced before office working hours, so allowing the Client to return to normal business.

The actual working programme developed for each of the four zones on every floor was a period of 14 weeks and this is briefly summarised as follows:

Programme prior to site installation works

- Release of Design Information.
- Site Surveys.
- Placing orders.
- Commence full Working Drawings.
- Issue of Working Drawings for Aproval.
- Manufacture of Ductwork.

Programme of site works

- Commencement of Work on Site.
- Removal of all Redundant Services.
- Extension of Ductwork to Offices.
- Extension of Pipework to and through Offices.
- Installation of Grilles.
- Installation of Fan Coil Units at Building Perimeter.
- Commissioning.

The various zones of the building were released to the contractors in various combination of one, two, three or all four zones in a floor, with this type of installation this could be achieved.

The commissioning of all these works was carried out with the installation on a progressive basis, both on the water and air volumes, and as each phase was concluded a recheck of the established figures was carried out and any adjustments made as necessary.

To enable the fans and systems to be gradually brought on line, motorised inlet guide vanes and, on the office floors acoustically treated constant volume boxes, were installed, the constant volume boxes on the first phase were capped off and were extended under phase 2.

The pipework on each zone was installed with regulating valves and a pump head relief system, again to allow gradual phased extension.

Upon conclusion of the installation works, a final inspection of all air volumes and water flow rates was made to ensure that during the phased installation the original settings had not changed.

Air conditioning system selection: *A.J. Mason*

Our terms of reference were to consider alternative methods of air conditioning to provide a good standard compatible with the expectations of a leading international company headquarters, paying due regard to the need to keep the Client effectively in residence whilst the refurbishment proceeds, capital and running costs, loss of useful office space and a need to maintain essential services in event of a power cut.

The study was to present the options in a form that would enable the Client to decide on the overall preference.

It is important to appreciate the type of building under consideration. This is of heavy construction with offices above ground (first to eighth floor), and main entrance hall, catering, workshops, boiler and refrigeration rooms, social club, conference rooms, communications etc. at ground floor level and below.

The office area totalled approximately 17,200m^2 with an average of 2,000m^2 per floor. To provide natural light and ventilation the building contains a number of internal lightwells.

To achieve the overall intent we listed the following aspects of paramount importance:

- Installation cost
- Installation time
- Running cost
- Minimum loss of existing office space
- Fresh air ventilation and extraction
- Individual temperature control
- Ability to install system with minimum disruption to occupant
- Noise
- Energy conservation

The existing single glazed windows were far from air tight and on their own by today's standards would have been considered unsatisfactory.

The impact of this upon capital and running cost was indicated in a schedule of running cost comparisons. Before a final decision was taken the cost effects of replacing existing windows, provision of anti-sun blinds, anti-solar glass and the aesthetic effect of replacing the windows, were all considered. As a separated exercise the capital and maintenance cost of the building works were assessed along with acoustic improvements to the front offices facing the Embankment.

As a result of this overall appraisal the provision of secondary internal glazing followed, this simplifying the construction problems and furthermore permitted the installation of mid pane anti-sun blinds.

With particular constraints of this building a long list of possible air conditioning systems was quickly narrowed down to two principle alternatives, each with variations around the basic concept. The principle alternatives considered were:

- Local individual room heater/coolers leaving ventilation by natural means.
- Mechanical supply and extract (heated/cooled) provided either as sole means of heating and cooling or supplementary to room units.

Under the first alternative we considered:

Split room air conditioning units with external air cooled condensers natural ventilation.

Large split packaged air conditioning units, without and with central fresh air plant.

Fan coil units served from central chillers/boilers, without and with central fresh air plant but retaining the existing air conditioning system for the areas presently served.

Under the second alternative we considered:

Dual duct VAV all air system, decentralised air handling plant.

Dual duct VAV all air system with large central air handling plant.

Fan coil units served from central chillers/boilers without and with central fresh air plant throughout the office accommodation.

The relative advantages and disadvantages were appraised against the following headings:

1 Installation costs
- Air conditioning
- Heating
- Electrical power
- Lighting
- Builders work

2 Running costs per annum
- Electricity
- Oil
- Maintenance

3 Area of lost office space

4 Areas to be vacated for construction and plant space required

5 Amenity value:
- Control of individual room temperature
- Fresh air ventilation
- Smoke extraction
- Noise
- Office disruption

These comparisons were presented in a table form

- covering the capital and running cost appraisal with reference to loss of useable office space and
- setting out an amenity value table.
- The effect of the systems of useable office space was presented in graphical form on a floor by floor basis.
- Outline programme periods.

Arising from the study and with the Client's own assessment of the importance of the various amenity factors a four-pipe fan coil system with primary fresh air and extraction was selected.

This decision having been taken there followed an intense period of Client consultation and design and from this evolved a requirement that the works would be effectively carried out in two parts.

The first part being the establishment of all essential new plant and distribution mains through four main riser ducts whilst the building remained in full use.

It is perhaps useful to mention that the building was served by horizontal shell type steam boiler plant which met all the heating needs with some spare capacity. Low pressure hot water radiators provided the basic heating via steam/hot water non-storage calorifiers and a small steam heated absorption chiller unit provided cooling for a partial air conditioning scheme (induction units) to the south facing rooms.

The second part involved the installation in the offices which was specified as being carried out in thirty-six stages over a period of approximately three years. The principle being that each stage represented approximately one quarter of a floor, three months allocated to complete a stage and a new stage was commenced each month. This meant that after the first three months, three stages were always under construction and in different degrees of completion. During this period the existing heating and limited air conditioning system had to be kept in service.

As each stage was completed it was returned to the Client for occupation. This principle meant that the central heating, cooling, air handling, and distribution of same had to be capable of operation initially under low load as the first stages were completed and progressively take up the load as works progressed. This was achieved centrally by:

- In the case of chilled water having a buffer storage vessel with primary and secondary circulation, and pump head relief.
- Heating with zone pumps and pump head relief.
- Air handling having duplicate fans and inlet guide vane control and zone constant volume boxes.

The allocation of plant rooms for the new plant meant that in principle the new plant had to be accommodated in spaces either released from low priority areas such as basement storage or in the case of the main primary air plant serving above ground offices a new plant room provided.

The pressure on space inevitably means that existing areas do not provide as much space as is desirable and this makes the need to properly plan and detail the layouts more critical.

For instance, Unilever have their own bore hole water supply which happens to be stored in underground tanks below the floor of the refrigeration plant room. This presented a difficult problem of supporting heavy plant above the tanks without placing a load on them.

In recognition of this potentially difficult aspect, and mindful of the delays that can follow if not fully considered in all aspects, Buckle & Partners were also commissioned to prepare the working drawings.

The problem of insufficient plant space at roof level, with the top floor offices having a good view over the Thames preventing space being found within the building, was solved by removing high level water storage tanks from a roof top enclosure and dispersing the storage into a series of small tanks on roof level covered over by a light weight structure. The plant space vacated was used to accommodate the primary air handling plant.

The establishment of vertical distribution risers complete with supply and extract air ducts and heating and chilled water with branch connections at each floor and all connected up to central plant in the first stage, provided the means and flexibility to progressively refurbish the accommodation under partial load of the main plant.

This degree of flexibility is fairly easily achievable with pump head relief but where a large variation is expected on fan air volumes, as in this case a 36:1 turn down, a twin fan assembly each with inlet guide vane control and distribution ducts having strategically placed static pressure control boxes were incorporated in the design.

Heat recovery in the form of run around coil at the central air handling plant and condenser cooling water from the central refrigeration plant was available to preheat fresh air intake and domestic hot water respectively.

The splitting up of the central chiller plant into mechanical compression and absorption type plant is a rather special circumstance brought about by a requirement to keep the building in use and operational in the event of a major power failure and also enable some control of maximum electrical demand to be exercised.

The electrical distribution network from the incoming supply point was split into three separate distribution systems titled principle, essential, and subordinate.

By this means a reduced power distribution can be easily controlled within the capacity of standby diesel generator plant (2 units @ 640kVA each). It is under this condition that only absorption refrigeration is run as it enables 50% of total cooling capacity to be provided for a reduced electrical input in relation to the turbo machines. The relation between electrically driven turbo compressors and absorption machines is 2 electrically driven vapour compression machines each at 880kW (250 tons), and 2 steam heated absorption chillers each at 914kW (260 tons) of refrigeration.

The condensers are water cooled, with forced draught cooling towers placed on the eighth floor roof. Special low speed fans were provided to keep noise levels to acceptable limits without attenuation.

Having provided the alternative type of chiller machine it places a requirement upon the plant engineer to operate the alternative plants under normal conditions in the most cost effective manner.

The cost of electricity to this building is based on an M.D. tariff with the essential electricity users in winter producing the maximum demand and consequently the tariff. If the air conditioning load can be satisfied from the absorption machines via the steam boilers the increase in M.D. as a consequence of providing cooling will be minimised, though the energy input to the absorption units when including steam requirements is greater than that of the electrically driven machines.

In this particular project the theoretical winter running costs are £2,556 per month absorption and £2,655 per month vapour compression.

The comparison between the options is extremely close, to the point where it will be influenced by future fuel costs. If the present trend of reducing fuel oil cost against a steady or possible increasing electrical charge is maintained, then the absorption machines would come more into use. Should the reverse apply in respect of fuel costs then the electrical driven machines would be viable for winter use.

There were in excess of 600 perimeter fan coil units installed and a careful evaluation was carried out to establish the most suitable in relation to low maintenance, low noise level, and good quality. The units were variable speed enabling the output to be adjusted to suit the need, with particular reference to low noise level. The design noise for this project was NR40 for general offices and NR45 for open plan offices.

To achieve economy of operation from the end user, control of the fan coil units during normal working hours is from a central station and at the end of the normal day all the fan coil units are switched off. However, for staff working late there is an extended run button which extends the running time for 1 hour and after this period a further operation of this extended run button is necessary.

Further limited local control of temperature is provided at each fan coil unit, limited to around $\pm 2°C$ about the set point. This set point can be altered centrally to suit changes in legislation and general space temperature as required without the need to adjust each controller of which there are in excess of six hundred.

Primary air to the offices was introduced through ceiling grilles and the extract grilles were likewise ceiling mounted. Fresh air was introduced at the rate of 12 litres/second/person.

The provision of 100% supply air in the primary supply system fully ducted extract air did not readily lend itself to heat recovery from light fittings. It was, however, practical to incorporate a run around coil in the fresh and exhaust air connections in the plant room which recovered approximately 50% of the rejected heating/cooling in the exhaust air.

As a final point I mention the close contact we had with all concerned on this project which to achieve all start and completion dates involved us in progressive design. The purchase of long delivery items was initiated by us through letters of intent following seeking alternative quotations, which was taken over by the Contractor as each section of the works was passed on to him.

Building management system. Integration with the building programme: *John D. Glisson*

Automatic temperature controls

Because of the seniority of the Head Office Building occupants, Unilever determined that the Fan Coil Unit configuration should provide each occupant with a 'Micro-Climate', with provision for adjustment of the temperature setpoint within limits determined remotely by the Building Maintenance Engineers. This was an attempt to give the occupant some freedom to choose his space temperature but keep the overall energy usage within reasonable limits.

Because of the large size, complex nature and phasing of the installation it was determined that the flexibility of pneumatic controls made their selection more appropriate. This was because of:

- Lower costs
- Ease by which complex over-ride functions can be achieved
- Readily extended and modified to meet the Client's changing needs during design and after occupancy.

On many an occasion during the installation, and since, this has proved to be sound judgement.

Building management systems (BMS)

The autonomous nature of the Control System dictated that the BMS approach was an overlay supervisory and monitoring system with the exception of using BMS to considerably reduce the cost of the remote adjustment of the fan control unit zone temperatures and central plant control functions.

Following a competitive tender involving three suppliers the BMS selected for the building was the Johnson JC/80/55. This was based upon a lowest total cost criteria against technical performance.

The JC/80/55 has a powerful suite of software programmes providing separate packages to cover the disciplines of HVAC, Fire and Security all running on a single mini-computer configuration. The particular strengths of the JC/80/55 at this time (1979) lay in its System Format and the extensive use of English language.

System Formating automatically groups individual points into a

'System' of related functions regardless of their location within the building. For example, all points associated with the Chiller Plant would be displayed on the VDU as a system even though the cooling tower points are connected to the roof-level outstation and the chilled water points are connected to a basement outstation.

English language is used to describe the points monitored by the system, for instance, a dirty filter alarm on the 8th floor HVAC plant is described on the System as:

HVAC 8TH FLR. PLANT FILTER DIRTY

The letters underlined are the point number by which the operator addresses the point via the keyboard.

Most other systems at that time used a Point Format and used a system of coding related to the location in an outstation to communicate with the operator.

There are operator locations, one for engineering functions, the other for security. Each station comprises a Visual Display Unit (VDU) Keyboard, Printer and Slide Projector.

Should, however, either of the operator locations become inoperative due to fire or flood for example, the system will automatically transfer the control of that function to the other location.

There are 16 outstations located strategically throughout the building, all connected to a common communications trunk in a loop configuration. All the various sensors and items of plant to be monitored or controlled by the BMS are connected to the outstations. There was 1,325 points on the system made up as follows:

270 Temperature Inputs
50 Remote Resets
450 Start/Stops
366 Alarms — M & E
60 Alarms — Fire Zones
55 Alarms — Security
74 Watchtours

The software features provided are:

Basic system

- Acknowledge
- Date/Time
- Item Log
- Lock/Unlock
- Lockout Summary
- Alarm Summary
- Off-Normal Summary
- Start/Stop Control
- Ignore/Execute
- Binary Reporting
- System Fault Advisories
- Command Error Messages
- Loop Advisories
- Date Display
- Operator Display Device
- Automatic Slide Select
- English Language
- System Format

H.V.A.C. Application software

- Analog Reporting and Limit Control
- Remote Setpoint Adjust
- Customer Defined Summaries
- Trend Logging
- Intercom Communication
- Password
- Programmed Start/Stop
- Special Programming Capabilities
- Auto Peripheral Transfer
- Software Interlocking
- Auto Lockout
- Remote Building Outstations
- Power Fail Motor Restart
- Remote Alarm Device
- Analog Point Totalisation
- Run Time Totalisation
- Maximum Demand Limiting
- User Programmable Data Base Editor (on-line update)
- Calculated Point

Fire and security application software

- Access/Secure Monitoring of Security Points
- Test/Reset of Remote Alarm Functions
- Security Guard Watchtour Programme
- Programmed Access/Secure
- Supervised Security Zone Monitoring
- Supervised Fire Zone Monitoring
- Ionization Fire Detector Monitoring

The Basic System Software provides the operator with a centralised alarm and plant status monitoring system with the facility to acknowledge and summarise current alarms. A printer is provided to record all operator activities, alarms and off-normal situations as they occur. The slide projector automatically selects a diagram depicting the layout of the plant relevant to the point in alarm.

The HVAC Application Software considerably increases the operator capability to remotely control the plant. For this reason a system of *Password* was introduced to prevent unauthorised people accidentally or intentionally compromising the System.

The Application Software includes for *Analog Monitoring* and control allowing the operator to remotely display, record and adjust temperature and humidity points. High and low limits can be set on these points so that alarms will occur when conditions are off limits.

The feature *Trend Logging* allows the operator to select a number of points to record on the printer over a period of time. These points may be temperatures or plant status and provide a powerful analytical tool for the plant engineer.

Programmed Start/Stop facilitates the automatic switching of plant and lights etc. to complex time patterns giving considerable savings in plant running and plant engineers time. The Systems will automatically record the run time of connected plant which is a considerable assistance in planning maintenance activities.

Software Interlocking gives savings in site and panel wiring costs as it provides the interlocking between items of plant such as Supply and Extract Fans within the computer software. It means that subsequent changes are easily achieved. We have not provided safety interlocks in software, these are achieved in the conventional manner through hard wiring.

The feature *Power Fail Motor Restart* will, upon restoration of electrical power after a shut-down, automatically restart the electrical services in order of priority and safety to prevent overloading of generators or circuits. This allows the building to get back to normal in the shortest possible time during a period of much confusion.

A group of *Energy Management Software* programmes are provided to automatically record and minimise the energy consumed by the plant. *Optimal Start/Stop* will delay the morning start up of the plant until the last possible moment, taking into account the prevailing external and internal conditions and the required occupancy time and temperature.

Similarly, optimal stop will advance the plant off time to take advantage of the scope for acceptable temperature fall or rise prior to vacating the building.

Load Cycling will at periods of low load switch off plant for short periods to reduce energy consumption whilst maintaining space conditions.

The *Maximum Demand Programme* monitors the MD meter and automatically switches off pre-determined electrical loads to minimise maximum demand surcharges and reduce energy consumption.

The *Fire and Security Application Software* is designed such that the computer serves both engineering and security functions.

The two activities remain separated so that engineers cannot interfere with security functions and vice-versa. The exceptions to this are fire alarms and lighting controls which affect both disciplines.

The System directly monitors the various security intruder detectors provided throughout the building. This is achieved through specialist *Security Point Modules* that in addition to monitoring the sensor, supervises the wiring to detect any fault or tampering. The *Programmed Access/Secure* function will automatically during periods of normal building occupancy inhibit any security alarms whilst still monitoring the status of detectures and field wiring.

The *Watchtour Programme* automatically tracks the progress of the Security Guards tour of the building by monitoring his operation of key-switches, strategically placed around the building. The programme allows for a varying pattern for each tour and times the Guards progress between tour points to raise an alarm should he be delayed or under duress.

The *Fire Alarm Programme* directly monitors the system of break-glass manual alarms which we had wired in a conventional manner to a fire zone panel. These alarms monitored by the JC/80 are interlocked in software to control a number of magnetic door holders furnished on the Smoke Doors.

Ionization Smoke Detectors located in the air-handling plant and corridors are directly connected to the outstation. This is achieved through specialist *Fire Point Modules* that in addition to monitoring the smoke detectors, supervise the wiring to sense any fault, provide the stabilised power supply to the detector and the remote reset function of the detector and the remote reset function of the detector. This System will on detection of fire simultaneously alert the security staff, reconfigure the air-handling plant to depressurise the fire zone and close the smoke doors.

Design

The design and integration of the BMS into the Project was unusually difficult. At the outset there was a limited definition of plant or points to be monitored. There was however previous experience of operating a monitoring and control system by Unilever.

Over the years leading up to the refurbishment, Unilever's engineering and maintenance facility had pioneered the installation of remote control and monitoring systems, in their various buildings in and around the City of London. This proved to be a cost effective method of supervision of plant and equipment installed at each location.

This had two important influences on the engineering decisions made on Unilever House.

- Unilever's Engineer was able to define his operational requirements based on sound experience of monitoring and control and,
- The experience gained from operating their own system had reduced the manning level in the Maintenance Department, which reflected the benefits of remote monitoring and plant supervision identified by current BMS Systems.

Identifying operational needs

Successful BMS Systems generally require an experienced engineer to ensure a correct interpretation of customer requirements.

In the Unilever House project our engineers were able to communicate directly with Unilever's engineer, which proved a considerable advantage when they were designing the anticipated future requirements of the building within the overall cost constraints. Unlike most contracts, we were directly employed by the occupier; the benefit of this is born out by the comment of the Unilever Engineer at the end of the contract when assessing the interpretation and current use of the building. We were not far out, were we.

To do this, much time was spent with Unilever's engineers to establish how they ran their operations and what management structures existed. This enabled us to select points to monitor and control that would directly benefit their operation.

During this period we sensed a slight anti-computer or "I'm not a typist" reaction to the System. This was not completely dispelled until the

System was installed and we were able to demonstrate the ease of operation. This was achieved through a planned training programme involving both factory and on-site courses for their engineers and was helped considerably by the plant coming on-line gradually giving the engineers, who were more set in their ways, time to adapt. Another ploy used, was to at first issue a low level password and thereby give the engineers the confidence that they could not do too much harm.

Two particular operational needs that were easily addressed by the BMS were:

- The ease by which time programmes could be over-ridden to suit the unpredictable working hours of Senior Management, and
- The changing Fire Regulations that required automatic release of smoke doors upon the detection of fire alarm conditions.

Practical installation considerations

The refurbishment and phased nature of the project had little impact upon our construction activities as the majority of plant was new and our work was scheduled to be in occupied areas. The sizing and routing of cable carrier systems however did require careful consideration to ensure capacity for future unknown requirements without excessive expenditure.

The outstations were installed as work progressed down the building and as the zones of pneumatic controls became available they were connected to the outstation and brought on-line. This progressive build-up of the system needed close co-operation with Unilever's Engineers and careful documentation of the acceptance test of individual points as they came on-line.

The 'On-line Update' feature of the JC/80/55 proved essential to this exercise as points could be brought on-line without the need to shut down the System and disrupting occupied areas.

Lessons learnt

For reasons stated earlier the BMS Supplier was directly employed by the Client. Johnson Controls had already secured the contract to provide the automatic control systems and motor control panels, this contract was made with the M & E Contractor.

As earlier discussed this had considerable benefits for the design of the BMS. We did, however, experience inefficiency in the manufacture of the motor control panels, these were placed on order in advance of the BMS and subsequently no allowance was made for the provision of remote monitoring and control, this lead to modifications and increased cost.

Similar circumstances apply to the control of lighting and fan coil units where the cost of such items as time switches and interlocking relays could have been saved.

The justification for the BMS was based upon reduced manpower and energy conservation. However, due to occupant resistance the predicted energy savings have not all been realised.

The identification of loads suitable for load shedding without causing occupant concern was extremely difficult and subsequently load shedding has not provided the anticipated savings.

The automatic reduction of lighting levels proved very unpopular and not implemented. Furthermore, the load cycling of the FCU's proved unacceptable. This is due mainly to the sudden reduction in noise drawing the occupants attention to the fact that his unit had turned off. This, despite space temperatures being correct, caused many complaints.

However significant savings are achieved by load cycling the primary plant which is not perceptible to the occupants.

The main benefit of the system has been in the areas of supervision and monitoring of plant and building safety, together with close time control of unoccupied areas. This has produced energy saving but more significantly enabled Unilever to absorb a fourfold increase in services and responsibilities without any increase in the numbers of engineering staff.

Such has been the success of the BMS project the system has been extended to the new North Wing of Unilever House and over telephone lines to United Africa House and Hesketh House, two satellite buildings.

10 Inmos Clean Room for VLSI Microchip Manufacture

Introduction: *Don Storton*

The semiconductor industry as it had developed over the years since Shockley and his associates first discovered the transistor in 1948, has constantly striven for a higher degree of miniaturisation. This drive became emphasised with the advent of the integrated circuit, invented by Jack Kilby at Texas Instruments in the 1950s. Later developments in the 1970s and 1980s known as LSI (large scale integration) followed by the latest techniques of VLSI (very large scale integration) have demanded geometry which is currently down to the 2 micron level. This means that the width of the insulating gap between two conductors on a VLSI chip is two-thousands of a millimeter.

In order to manufacture devices with such minute patterns, is has been necessary for the industry to develop clean rooms of extreme cleanliness and very close environmental control. Particle control was the most serious problem to be overcome, as the normal human body is constantly shedding particles of dead skin known as epidermal scale which are sized from 0.3 μm upwards. Furthermore normal atmospheric dust without the most careful filtration would also be a killer to VLSI devices.

The method of control used in clean rooms is now universally VLF (vertical laminar flow). This consists of constantly circulating air over the full area of the working space at a velocity of 100 ft/min. through a so called HEPA filter thus ensuring that any particles generated within the space are swept down and out of the clean room.

These clean room specifications were formulated in the United States and are laid down in US Federal Standard 209B *Clean Room and Work Station Requirements, Controlled Environment*. The part of this specification which is now regarded as mandatory by all VLSI manufacturers is the class 100 which requires particle counts not to exceed 100 particles/cu ft of size 0.5 microns and greater.

The very latest technologies require so called class 10 clean rooms which although not yet recognised by Federal Standard 209B, requires particle counts not exceeding 10 particles per cubic foot of size 0.12 microns and larger.

The Inmos clean room

When the design of the Inmos clean room or Fab as it is known was first discussed in 1978, it was specified that the latest technology was essential as the geometrics to be used were among the smallest possible at that time. The concept employed was the service chase design in which conditioned HEPA filtered air was introduced into a chase which carried all pipe services used in manufacture of wafers. This is then drawn through HEPA filtered VLF hoods mounted over the fabrication benches in the clean room and discharged downwards at 100 ft/min to continuously wash the work surface with clean air. The air is then returned under the bottom of the service chase wall, which is arranged to form a 150mm gap between wall and floor, and thus back to be recirculated through the VLF hoods.

Semiconductor manufacture requires a considerable amount of chemical extract from wet etching, furnacing and other chemical processing. This extract is removed to scrubbers and then discharged to atmosphere. The make up air from the main FAB air handlers must therefore make up this volume of air as well as providing excess to maintain positive pressurisation of the whole clean room.

This clean room concept was first tried out in the Inmos Colorado Springs Plant and found to be very successful. It was subsequently refined in design for the Newport plant of Inmos Limited and takes the form of a clean room containing 36 chases and bounded by the service or line maintenance areas.

The clean room is some 3000 square meters in area and is fed with conditioned HEPA filtered air from 3 main AHUs each handling 33,000 cfm. Chemical extract amounts to approximately 40,000 cfm so allowing for pressurisation and leakage approximately 55,000 cfm is returned to the AHUs.

Specified conditions for the most sensitive areas of the clean room are temperature 21 °C ± 1 °C and humidity 40% ± 2½%. The remainder of the room requires 21 °C ± 2°C and 40 ± 5% RH.

The purpose of the main AHUs besides HEPA filtering the supply air is to control the humidity condition of this air so that room humidity is controlled to the required limits. The unit therefore comprises mixing box, bag filters, frost protection preheater coil, chilled water coil, fan, HEPA filtration and finally steam humidifier.

In practice, it is found that an off coil temperature of about 8 °C will give the correct humidity in the clean room and this temperature gets increased by approximately 2°C by the heat generation from the fan motor. Room sensors then trim the humidity to a level which will result in the appropriate 40% RH room condition by means of a steam humidifier. Air at about 10 °C is passed down into the Clean Room via the external supply ducts and this is then distributed to each service chase via a constant volume box and reheat battery. The reheat battery, fed with LPHW, is controlled from an individual thermostat in each bay of the clean room. After discharge into the chase the air is picked up by the VLF units which comprise prefilter, fans, HEPA filters and lights and is then discharged down over the work surface.

Each service chase receives about 2800 cfm of make up air compared with about 14,400 cfm of air passing through its VLF hoods. There is thus about 5 times as much VLF circulating air as conditioned air from the main AHUs.

This method of clean room construction succeeds in producing class 10 conditions at the workbench, class 100 in the bays, and slightly higher than class 100 in the centre aisle of the FAB.

Cleanliness at this level is difficult to comprehend but some idea is given by the fact that the Inmos clean room is over 10 times cleaner than a hospital operating theatre.

Main service supplies

In addition to the clean room the whole 9,000m^2 Inmos Newport building is fully air conditioned by the methods outlined later in this chapter.

The supplies for this whole system come from a chilled water circuit and a hot water circuit carried to each of the air handling units mounted externally above the central building spine on two levels. The chilled and hot water pipework runs from a mechanical plant room of some 500m^2 along the spine at high level and then through the roof to the AHUs at each location.

The plant room contains 2 low pressure hot water Hoval boilers each of 5M BTU/hr capacity, gas fired, and producing water at a temperature of 85°C with duplicate HW circulating pumps. Each boiler also incorporates a domestic hot water calorifier supplying hot water to cafeteria and toilets.

Also in the plant room are three 400 ton York Centrifugal Chillers of the open drive type, producing chilled water at 4.5°C. The condenser water from these is fed to three Baltimore induction type cooling towers mounted on the roof above the west end of the spine. Automatic water treatment is used for chilled, condenser and hot water circuits.

Chilled water is powered around the system by three chilled water pumps feeding a common header which then feeds the three chillers. This enables two pumps to be used with only one chiller in winter to maintain circuit pressures at the AHUs.

Also contained in the plant room are manufacturing services including air compressors, drier, vacuum pump and a very large deionised water plant producing high purity water at >18 meg ohm/cm. The electrical substation containing three 1.6 Mva 11Kv/415v transformers and switchgear completes the plant room layout.

Controls

The original Air Conditioning installation employed Johnson pneumatic controls operating on three way valves for both chilled and hot water in each individual AHU. Each main AHU consists of a supply unit and a separate return air unit. This return air is mixed with fresh air at the entry to the supply unit and the proportions were originally controlled by a total enthalpy controller operating variable vanes on both fresh and recirculation air. The concept of this was to economise on chilled water by using ambient cooling in winter and also shutting down fresh in summer if external air exceeded return air temperatures. However due to the changes in manufacturing methods since the original design of the building, chemical extract has increased to such an extent that 45% of the incoming air must be make up fresh, irrespective of outside conditions, so the enthalpy control is no longer used.

Control of the AHU now consists of a thermostat set to maintain a duct temperature of 10°C in the output duct, controlling the chilled water three way valve to maintain this duct temperature. In winter conditions this duct stat also controls the hot water valve to give frost protection. There is also a separate freeze protection stat which cuts out the AHU if temperatures of below 2°C are seen in the space between heating and cooling coils.

Each area of the clean room controlled by a main AHU has two humidity sensors whose outputs are averaged then fed back to the humidifier in the appropriate unit. The humidifier is a nine step steam producer with a total capacity of 90kW.

The control principle therefore is that the chilled coil will wring out moisture to below the point which will result in the required 40% RH in the clean room, and then the steam humidifier adds moisture to trim the clean room conditions to specifications.

In each individual bay of the clean room a thermostat on the outlet grille for return air senses the room temperature condition and controls the three way hot water valve to its corresponding reheat battery thus trimming the inlet air temp to maintain temperature specifications.

Subsequent additional installations provided since the original contract have used Staefa electronic controls in place of pneumatics. These have proved much more stable in controlling the most sensitive area requiring ± 2½% RH and ± 1°C. Pneumatics will control to these limits but tend to drift whereas the electronics do not. All future installations will therefore be electronic.

Capital costs

The total capital cost of the Newport Fab 2 building in 1981 was £10M. Of this the total air conditioning cost was £2.8M excluding civil works in connection with these services. Electrical costs are also excluded.

This Air Conditioning cost is for the whole building but an estimate of the clean room costs alone is approximately 75% of this total, equating a cost of £672/m^2 of clean room space.

Future developments

This clean room concept has proved very successful for the geometries currently in use by Inmos, but increasing miniaturisation in semiconductors will make new demands on manufacturing environments. A future clean room would require a full VLF ceiling with through floor air return system. At Inmos we have already gone towards this concept by converting certain areas of the clean room to 100% VLF and this has proved very successful with the side wall return air concept which is satisfactory for a space of width not exceeding 4½M. However new clean rooms will be complete open spaces on a suspended floor with full plenum to supply the HEPA filtered ceiling above the work space and a perforated raised floor to the clean room passing air down to a full height floor underneath which will be the return air void. Conditioning of the air will take place in side towers spanning between the lower return air void floor and the upper plenum space. In these towers will be prefilters, chilled water coils, and large axial flow fans to move the enormous volumes of air involved. Pipe services, electrics and drainage will be run in the lower floor and pass up to the production floor as required.

Fresh air make up will be conditioned in a separate unit and then fed into the ceiling void to be mixed with the recirculation air.

The costs involved in this type of room will be an order of magnitude higher than the current Inmos type room, but increasing miniaturisation in semiconductors demands that these costs will have to be met to achieve acceptable manufacturing yields in the future.

High tech and visual aspects: *Colin P. Dobson*

Inmos with its striking external ductwork looks and is a building for services. The services account for well over half of the total building cost.

As with many recent Semi-Conductor plants, the Design Team lead had to be undertaken by either the Building Services Engineers or an inspired Architect who sympathises with and understands building services. Richard Rogers and Partners Ltd are one of the few Architects who appreciate services and uses them as part of the building form to maximise the internal space.

The original concept drawings for Inmos were produced by the Architects in 1979. As you can see they were very striking and controversial Figure 10.1. However, with input of the building services engineer and, a little help from the Newport Planning Dept., only the air related systems were retained on the spine, the wet and gaseous systems being located in either the internal power house or external tank farm. The services tunnel was also dropped at this time.

The building is fully air conditioned and consists of five different types of area each requiring different types of air conditioning systems. These are:

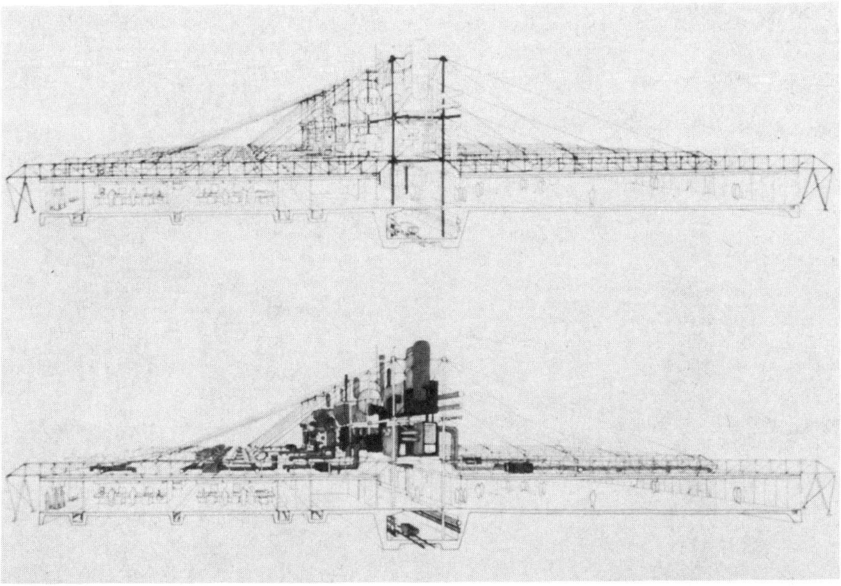

Figure 10.1 *Architects drawing*

- Cleanroom 3,000 m^2 All air terminal reheat
- Offices 1,500 m^2 VAV System
- Laboratories 1,000 m^2 All air partial VAV with reheat
- Restaurant 500 m^2 All air system
- Central Mall 750 m^2 Four pipe fan coil units

It must be remembered that the design for Inmos started in 1979, some six years ago now, and the services design was based on the then current cleanroom conditions and accepted standards. Six years in the Semi-Conductor Industry has seen many changes and the basic design is now somewhat dated. Inmos staff are, however, continuously changing and updating services since the building was completed to keep pace with an ever changing industry. The basic services network and architectural concept has accepted these changes without major alterations.

The cleanroom (Figure 10.2) is the heart of the Semi-Conductor plant. All production takes place within this area, which is subject to stringent standards and conditions, and which must be maintained to enable the product to be successfully manufactured.

The standards which govern the design, construction and use of

Figure 10.2 *Cleanroom*

cleanroom is the British Standards Institute Publication No. BS 5295 or the USA Federal Standard 209B.

The environmental cleanliness is stated in terms of size and maximum number of airborne particles. When it is considered that one cubic metre of fresh air contains between 10^7 and 10^{11} particles of 0.5 micron size and above, some understanding of the following standards can be appreciated.

The Inmos cleanroom obtains the conditions required by Class I to BS 5295 and Class 100 to FED 209B which state that the particle count shall not exceed a total of 3000 particles per cubic metre or 100 particles per cubic foot of a size 0.5 micron or greater.

The cleanroom is also required to be maintained at 21°C ± 1°C and 42.5% RH ± 2½% RH. These conditions are required at all times as it is usual for cleanrooms to operate a 3 shift or a compressed workweek system which means that the cleanroom is always in production and therefore little time is available for shut down and maintenance, etc.

The cleanroom is served by three primary air supply units situated on the spine each supplying 15.5 m^3/s of primary air which provides filtered, cooled, and dehumidified air in the summer and filtered, cooled and humidified air in the winter.

The arrangement of a typical cleanroom air system is such that the return fan is located on the upper level of the spine whilst the supply unit normally heavier is located on the lower level. The supply unit consists of a mixing section (min. 50% fresh air) roughing and prefilter section, preheat coil, chilled water cooling coil (14 Row), supply fan with inlet guide vanes, discharge silencer, HEPA filter bank and steam injector humidifier. The casing of the unit was extended to accommodate a 1.2m wide enclosed service and maintenance corridor in which was also located a starter and control panel, steam generator and chilled and heating pipework with isolation and control valves. Access is also provided from this corridor through a hatch to the return fan above.

In order to maintain a relative humidity of 42.5% with a conventional water chilling plant producing chilled water at 4.4°C (5.5°C at the farthest unit) it was necessary to use 14 rows of Cooling Coil. Considerable moisture was taken out by the Cooling Coil due to the high fresh air content. It is now often a requirement of cleanrooms to have a lower Relative Humidity in which case a brine cooling system would be required or local DX or rotary air dryers.

The primary air at a constant condition of 10°C is reheated and supplied into each service chase, which are located between back-to-back work stations.

The primary air is mixed with room air in the proportion of 4 to 1 in the vertical lamina flow units. The mixed air is then passed over the final HEPA filter below which all production takes place.

The office areas of the building are air conditioned using an all air variable volume system with perimeter heating units. To accommodate a higher ceiling it was necessary to locate the supply ductwork partially below the open type grid ceiling. Variable volume supply outlets on the side of these ducts were located 75mm below the open type ceiling. Tests undertaken had proved that a cool air stream 75mm below the open ceiling was unaffected by the vertical fins, furthermore the variable volume valve, which had a constant volume bypass at the bottom, provided the throw required at full and partial air flow. Thermal insulation and cladding to the spigot air valve and outlet grille provided a neat flat oval connection to the circular branch duct.

Perimeter heating was provided by mini fan coil units located at high level above the open ceiling operating on a temperature and solar compensated hot water circuit.

The Laboratories also had a separate supply unit and exhaust fan but because of a future requirement for major exhausts from this area, only a partial VAV system was used: Terminal boxes, modulated down to half volume before a hot water reheat coil provided the necessary control. Circular diffusers were used to distribute the air in this mainly internal area. Perimeter mini fan coil units were again used to offset the perimeter heat loss.

A mainly fresh air system served the Restaurant area providing make-up air for the adjacent kitchen. Four large constant volume zones each with a reheat coil provided temperature control. The ductwork had a similar appearance to the office areas, a small amount of air was returned to the supply unit but the majority was exhausted to atmosphere via the kitchen exhaust canopy.

The central mall which is the main circulation route was heated and cooled by four pipe fan coil units located in adjacent areas. The fan coil units had a fresh air system which also served several areas adjacent to the mall such as Locker Room, Security Office, etc.

The architectural integration of the services into the external building form is probably the most exciting and unusual feature of the building.

The main reason for the external services is to reduce the building envelope. It is estimated that if all the plant was accommodated inside the building, apart from that normally housed outside, the building envelope would have to be increased by 25%.

The roof plan (Figure 10.3) shows the layout of the various supply and return air units, fume exhaust scrubbers and cooling towers which were originally installed. As you can see very little space was left. Inmos has, however, installed additional equipment to suit recent modifications, which the building has accepted. Not many internal plantrooms could accommodate this type of change.

The ductwork system on the roof is the main horizontal distribution zone conveying air to all parts of the building. The ductwork passes through primary and secondary steelwork trusses on a common centre line. In order that one duct system passes another without unsightly offsets, a cross-over box was devised. This become a feature of the roof ductwork. A typical sample is shown in Figure 10.4. The supply ducts change section but maintain velocity, whilst the low velocity return duct uses the balance of the space inside the box.

Considerable effort was put into the method of supporting the ductwork, both in the vertical connections to the air plant and to suit the 13.5m × 6m structural grid. The 450mm spiral ductwork was supported at 6 metre centres.

All supply and exhaust ductwork was fully pressure tested to 1500 pa. In

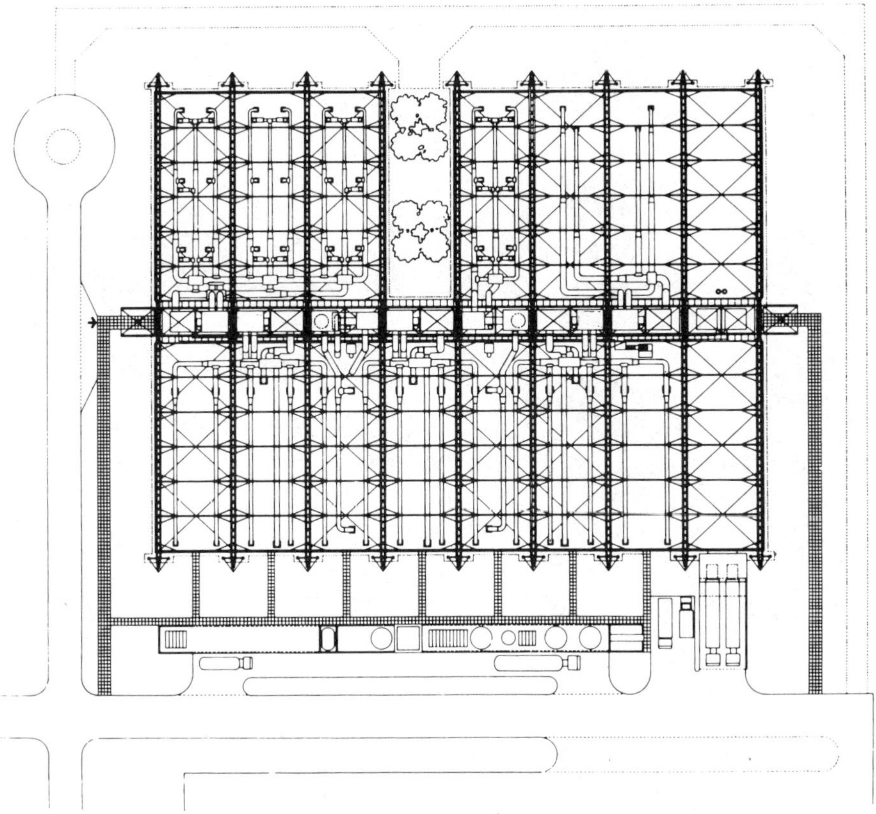

Figure 10.3 *Roof plan*

addition, all cleanroom ductwork was cleaned and sterilised in situ before systems were run up.

All external supply and return ductwork was insulated with 50mm high density 65 kg/m^3 insulation incorporating a vapour barrier and the whole clad with aluminium sheet finished in PVF2 coating.

People often ask me which was the most difficult aspect of this project: to give Inmos the conditions that they required to manufacture their product or to work with the Architects to achieve a structural and services integrated building; obviously it was the latter.

Figure 10.4 *Roof ductwork*

The design of a flexible production machine: *Mike Davies*

Context

At the time of the formation of Inmos, Silicon Valley in California was producing literally millions of microchips. A study of the industry processes and usage revealed that the servicing fabric of those microchip production plants was being continually modified to keep pace with new production technology which was emerging at an incredible rate. Not only were the production services and equipment being ripped out and replaced on a regular basis but major elements of the production buildings were being demolished and rebuilt to suit new technologies and layouts. Each time any major modification to structure or clean room layout was made, production processes which were hypersensitive to dust particle pollution had to stop and restart later with significant cost impact and technical yield and quality control problems.

In short the industry was characterised by continuous and rapid modification of building and plant, to the detriment of research, production continuity and with continuous management and planning headaches.

Further, the industry was bedevilled by production stoppages due to fires in specialised ductwork and servicing systems and by dubious safety conditions in technical areas.

Design concept

The evolution of the design and production concepts for the Inmos facility were evolved as a direct response to this unsatisfactory state of affairs which effectively served as a brief for what not to do.

It became clear that the structure and the plan form of any new Inmos building must not act as a constraint on microchip production, nor on building growth and continuous necessary modification.

Inmos themselves emphasised this point in stating that (a) they would be designing the production requirements and layout whilst we designed the building, (b) that they would change their mind continuously as they went along, and (c) that we would not be told what site the building was on until the design concept was established — in short — an architect's nightmare and a classic recipe for cost and time overruns. Flexibility was definitely required in the design process as well as in the building itself!

A number of basic design parameters were evolved by the Design Team and the Client.

1 The building concept should be applicable to any green field site.
2 The design and contract procedures should be capable of responding to a changing brief.
3 Rapid construction programme requirement: 1 year maximum from start of construction to clean room-handover to client.
4 Maximum internal layout flexibility.
5 Maximum services flexibility.
6 Clean room services access and modifications outside clean areas.
7 Column free clean room area.
8 Vibration free clean room area.
9 Capability for further growth of the building concept on the site.

Based upon these parameters, the design concept evolved with the participation of the client, from four basic decisions:

1 Adoption of a lineally expandable, lateral wing and central spine plan form.
2 The decision to modularise and distribute air handling plant rather than create a central air plant room.
3 Placement of modular air handling units above the centre of the building, avoiding long service runs from and avoiding the planning constraints of edge sited plant at ground level.
4 Adoption of long span column free wings with spans assisted from the central plant support spine.

The combination of these design parameters and initial decisions yielded a modular super-structure concept with a strong 'spine', and lightweight lateral suspended wings with large column free areas, with the building being capable of extension by linear addition of further bays.

The spine provides the main personnel entry and movement route at ground level, with main fluids and power distribution and air services platforms at high level and the lateral wings provide the enclosures for the clean room area and associated shipping and receiving zone on one side of the spine and offices and laboratories for assembly, testing and R + D activities on the other. The very compact fluids plantroom is placed alongside the laboratories on the opposite side of the spine to the clean room which allows future linear expansion of clean area activities. The cleanroom itself is sited under the north wing of the external envelope of the building above which main air supply, return and fume extract ductwork for clean areas run, utilising the external structural truss zone as a natural service run and leaving a completely free span production zone uncluttered by structural beams or by main air distribution systems.

Clean room design

Clean rooms are invariably associated with vast quantities of air ducting and with large numbers of piped and trunked specialised services.

The existing industry was studied and the two traditional solutions to clean room services access, high level or subfloor distribution, were each found to have particular problems.

In Silicon Valley and most chip production facilities, above-ceiling service distribution caused regular stoppages of production due to dust disturbance from human access and also carries considerable risk of fire in service voids.

Only by moving to full filter ceiling systems can the access pollution associated with high level distribution be entirely eliminated.

The sub floor distribution of pipe services automatically puts the clean room on a suspended slab with inherent cost and vibration problems and creates the problem of connecting through the slab with either in situ coring or service penetrations via pre-fixed sleeves which inevitably are not in the right place. Both end up polluting the clean zoneroom environment with dust and local vibration.

The Design Team and the Client decided to develop a side-access services system which avoided the problems of the traditional 'above or below' approaches.

By adopting this side servicing approach, no clean room services at all, apart from general protection sprinklers and fire detectors run above the clean room inner ceiling which is a sealed continuous white metallic skin 90 metres long by 22 metres wide.

Services distribution of up to 16 different services is achieved by high level side running principal pipework and trunking loops with sub loops or T's feeding a series of walk-in service chases. Each sub-loop services 16 metres of potential work station. Gas pipework supply loops for each service finger were prefabricated, tested and pipes nitrogen filled and sealed off-site and then brought to site and installed very rapidly.

To increase flexibility and aid interchangebility of clean room chase elements, and equipment and to eliminate superstructure borne vibration, all divisions, wall surfaces and structures within the clean room are designed as free standing modular demountable frame and panel systems with no connection to the ceiling. All equipment fixes back to framing rails which are clad with interchangeable wall panels only 2.5mm thick. These panels provide an air barrier and dust free surface as a division between ultra clean areas and the service chases and back up zones.

All clean room equipment sits on a continuous sealed vinyl floor on slab on grade which is vibration independent of the superstructure of the building. Specialised isolated slabs for sensitive equipment are also utilised within parts of the clean room area.

The slab also provides two collector trenches for wet services. These trenches and also air distribution and fume extract points are grouped outside the main clean room areas in fully accessible service aisles within which all back-up equipment is housed.

By these means over a half kilometer of potential work station run can be developed which forms an extremely flexible infrastructure for the clean-room and which can be grouped and changed at will, complete service chases being interchanged with no structural constraints or compromise to other areas of the clean room.

All internal divisions within the building are designed for flexibility; all partitions are demountable and even blockwork walls are all stack-bonded to allow new openings to be formed with ease through heavy mass walls. Inmos have made considerable technical changes and additions within the building as their needs evolve in clean room production, and general servicing and layout terms and the building has proved very accommodating to modification and change, including the creation of 5000 square feet of extra laboratory space by infilling a courtyard bay, with no significant disruption to other activities.

Both the overall building plan form and the clean room areas fulfill the need for high performance flexible usage and the nature of the structure and its typical cross section allow virtually any service to be brought to any part of the production and R + D areas of the building. Whilst the clean room is fully enclosed by a highly insulative building envelope, the offices, restaurant and laboratories, on the other side of the spine, look South, out onto woodland. These activities are all linked to the generous central spine which is virtually an internal top lit street which performs the role of visual security control, as well as being the principal reception, access and social space in a high performance yet friendly production building.

11 The Hongkong and Shanghai Banking Corporation New Headquarters, Hong Kong

Overview: *Derek Tuddenham*

By any standards the Hongkong Bank Group is a major international financial institution. As the primary note issuing bank and largest banking presence in Hong Kong it holds a special position in the Territory's commercial life.

It has also established a tradition of commissioning state of the art buildings for its Headquarters ever since it first acquired the site at No 1 Queen's Road Central in 1865. This tradition is exemplified by the 1935 Headquarters which, in its day, was the tallest building between Cairo and San Francisco and the first major air conditioned building in Hong Kong. However, by the late 1970s the Bank had outgrown this building which was dwarfed by its neighbours and was clearly unsuited to meet the needs of new banking technology in one of the world's largest financial centres.

A new Headquarters building was required and the Client intended it from the outset to be the best bank building in the world — an emphatic statement of faith in Hong Kong and the Bank's own future.

The agreed plot ratio for the site was 15:1 thus dictating a high rise block in a constricted urban position — traditionally one of the most difficult formats for both architect and engineer. The commercial district of Hong Kong already has an abundance of silver and gold curtain walled giants and executing a dramatically different building under these conditions required an imaginative brief and the Client's unstinting support throughout its interpretation. This the team had in full measure and at an early stage a number of key requirements were established for the building which characterised the visual form and development of the building services.

These key requirements are summarised as follows:

1 'State of the Art' Building Technology.
2 To build in Hong Kong to the highest European/North American standards.
3 A high ratio of net usable area to gross area.
4 Flexibility of internal planning and services capable of responding to a rapidly evolving banking technology and high departmental mobility.
5 A building and services design life of 50/25 years respectively.
6 Maximising the potential of the sites protected views of Hong Kong Harbour and Victoria Peak.
7 Building completion by 1985.

The Architect's response was a 46 storey steel frame building in the high-tech idiom which made extensive use of pre-fabrication techniques to achieve unprecedented standards of quality control within the specified construction period of 4 years.

The general statistics of the building are set out in Table 11.1. It is notable that the economies in services space associated with factory production techniques made a major contribution to the high ratio of net usable to gross area which was achieved.

Table 11.1 *Building statistics — general*

Gross Area	100,000m ^{2}
Nett Area (usable)	70,000m ^{2}
Height	179m
Floors above ground	46
Floors below ground	4
Floor Area - Largest (3 bay)	3,200m ^{2} (2,650m ^{2} nett)
Floor Area - Smallest (1 bay)	976m ^{2} (687m ^{2} nett)
Atrium Height	52 metres
Glazing	88% of principal facade
Installed Central Chiller Capacity	5 x 3125 kW (1 standby)
Installed Heat Pump Capacity	2 x 3750 kW (1 standby)
Seawater Pump's Capacity	1m ^{3}/sec
Incoming Power Supply	21 MVA
Installed Standby Generation Capacity	6 MVA (gas turbines)
Construction Programme begun	November 1981
1st phase occupation	July 1985
Occupation complete	November 1985
Cost	£500m (approx)

General design response

Following the appointment of the Professional Team in January 1980, the Client agreed to a 12-month concept development period during which the Architect's imaginative scheme was harnessed to the evolving details of the Client's Brief.

Throughout this period the Team embraced a whole building philosophy with the design process being informed by the needs of all disciplines and the construction methodology.

Key features of the design response were:

1 Research and analysis of all relevant systems and suppliers.
2 Appointment of a management contractor to provide close control of sub contracts and suppliers and advise on buildability.
3 A fast track programme to meet the Client's short construction programme.
4 Product development through working with industry to implement the results of the Team's researchers.
5 Off-site prefabrication to achieve "factory" quality in all building components.

The resulting scheme, approved by the Client's Board in January 1981, retained almost all the major features of the original competition scheme now fully supported by detailed engineering and focussed on the Client's confirmed requirements.

The general layout of the building (Figure 11.1) featured:

- Four levels of basement to house the Client's security vaults and main engineering plant.
- An open plaza at ground floor dedicated to public usage.
- A main banking hall set beneath a 10-storey Atrium.
- General office floors in the superstructure having large open areas unrestricted by a central core and intermixed with double height spaces for the Client's special facilities such as dining, function rooms and recreational areas.

At the top of the building is featured a helipad beneath which the Bank's skysign wraps round the E & M systems tank rooms.

General arrangement and scope of engineering services

In keeping with the concept of a light open superstructure, the central plant (chillers, pumps, transformers, switchgear, main water storage) is located in the Basement levels, and the office floor air handling units and other local plant are stacked outboard of the steel masts on the extreme East and West of the building in prefabricated modules. The gas turbine standby generators and domestic hot water boilers were also prepackaged in modules and positioned on the West side of the highest floors. Thus was avoided the encroachment on usable floor space of flues rising from plant located at Basement level.

Hong Kong has in the past often suffered fresh water shortages and seawater is frequently used for flushing. After a detailed economic study seawater was also selected as the means for heating and cooling the building.

To bring seawater to the building's basement, a tunnel was driven 350 metres into the site from new intakes and discharge facilities adjacent Hong Kong's famous Star Ferry Terminal. The seawater system pump house was located in the vertical pipeshaft near the intake.

As might be expected, the new Hongkong Bank features all the services associated with a forward looking international financial organisation. However, uppermost in the designer's mind was the knowledge that the systems and spatial allocations must take account not only of existing technology but also of changes in office technology yet to come.

The creation of sufficient flexibility to last the life of the building was the major driving force behind detailed design of the air conditioning, lighting, power and data communications systems. Flexibility permeates every aspect of this work, not only in the adoption of a raised floor for distribution throughout the offices but also in the total plug compatible, mix and match approach which enables lighting levels, work station services and air conditioning to be tuned to any department's or individual's needs.

With flexibility comes complexity. The Designer's appreciated that the owner occupier must be able to capitalise on the buildings flexibility without recourse to outside specialists each time, say, a major department was rearranged. Wherever possible, simplifying rules were set down for matching systems configurations to the particular user requirements. In the case of air conditioning, steps were also taken to design out much of the rebalancing associated with changed layouts.

Project organisation and overall programme

At an early point it was agreed with the Client that this ambitious project

Figure 11.1 *North–South section through building*

would best be served by organising construction under a management contract. Following an international competition, the Management Contractor was appointed approximately nine months into the concept development period (September 1980) and he then supported the London based design team in their final design submission and through the early months of the design detailing working under the Architect as design manager.

With the commencement of work on site in November 1981, the Management Contractor moved into his principal role as construction controller and co-ordinator in Hong Kong. Normal management contract rules applied, in that all work was undertaken by sub-contractors in contract with the Management Contractor who was not himself permitted to undertake any work directly.

After demolition and foundation work, steel frame erection began in earnest in January 1983 and the entire Professional Team transferred to Hong Kong to continue the design of work in the later sub-contracts and to provide on-site design liaison.

Figure 11.2 shows the principal elements of the project organisation at this time with a Project Manager now appointed to ensure tight cost control and the entire team answering to the Client's board sub-committee which regularly reviewed cost and programme and guaranteed the consi-

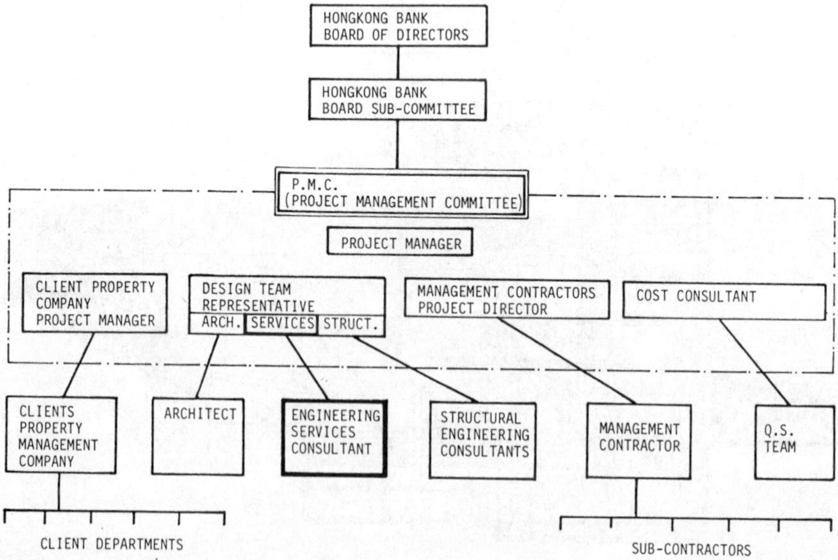

Figure 11.2 *Hongkongbank project organisation*

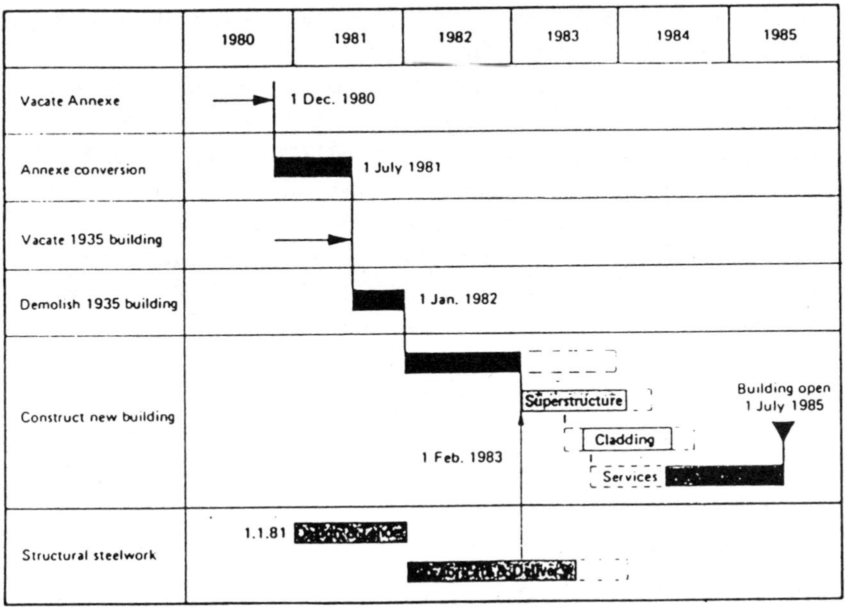

Figure 11.3 *Milestone programme (May 1980)*

derable Client input so necessary to a successful project of this nature.

Figure 11.3 shows the project programme as it was conceived in May 1980. Vitually every date was met within a few days and the Client took occupation of the first three zones of the building on 1 July 1985.

Energy considerations and building envelope analysis

As one might expect, energy optimization was an important concern shared by Designers and Client.

The Architect had always envisaged extensive use of glass and reconciling this with satisfactory air conditioning energy usage required particular attention.

It was clear that effective external shading was a fundamental prerequisite. Various configurations of shading and glazing were analysed during the concept development period using a modified form of the CIBSE/BRE admittance method on an in-house computer. One interesting result arising from the use of efficient external shading was the need to define a high diffuse design condition in the absence of direct insolation. It emerged that this design figure was approximately twice that obtaining under conditions of maximum direct solar gain.

The datum used for the comparison of alternative envelope configurations was a building of the same height and gross area arranged in the popular podium and tower format. Figure 11.4 shows typical results derived from the concept development period and compares this with the final scheme described later in this chapter by Ratcliffe.

Computer calculations of energy consumption were repeated at frequent periods throughout the design evolution. Table 11.2 gives a breakdown of the energy forecast for the final building.

Table 11.2 *Final annual energy forecast*

Electrical Energy	KW Hours x 10^6	Percent of Total
Refrigeration cooling and heat pumps	5.50	16.8
Pumping include seawater intake	5.23	16.0
Basement AC	0.90	2.7
Fans for AC and ventilation	5.46	16.7
Computer suite cooling	2.19	6.7
Sub Total for AC and associated equipment	19.28	59.0
EDP.	4.40	13.5
Small Power	1.05	3.2
Lighting include circulation	5.28	16.0
Signs, decorative lighting and accommodation	0.50	1.5
Kitchens	0.25	0.8
Docuemnt Handling	0.02	0.1
Seawater Chlorination	0.23	0.7
Lifts + Escalators	1.20	3.6
Water boosting, drainage pumping + DHWS pumping	0.14	0.4
Refuse System	0.30	1.0
TOTALS	32.65	100%

BUILDING FORMAT	REF	ENVELOPE CONFIGURATION	ANNUAL ENERGY CONSUMPTION AS % OF REFERENCE BUILDING	SHADING COEFFICIENT	CALCULATED DRY RESULTANT TEMP 1.5m FROM GLASS
PODIUM/TOWER TYPICAL DEVELOPERS FORMAT (USED AS REFERENCE) 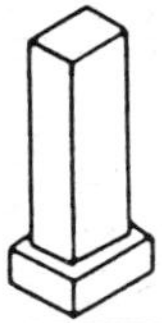	A	Podium/Tower configuration giving equivalent gross floor area to final concept development. 40% glazed on all faces with heat absorbing glass.	100%	-	—
HONGKONG BANK CONCEPT DEVELOPMENT 3-TIERED TOWER WITH EXTERNAL SHADING 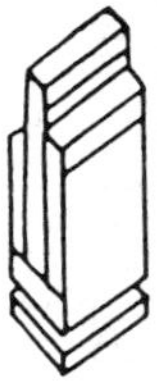	B	88% clear single glazing.	91%	1.00	30.0°C
	C	88% heat absorbing single glazing.	86%	0.50	26.0°C
	D	88% clear single glazing with closed reflective internal blinds.	85%	0.38	26.0°C
	E	88% silver reflective single glazing.	81%	0.26	25.5°C
	F	44% heat absorbing glass. 44% translucent insulated panels.	85%	0.34	26.0°C
	FINAL	88% double glazing with 10mm clear outer pane, grey tinted inner pane and perforated reflective mid-pane blinds	86%	0.34	25.0°c (with perimeter supply in operation)

Figure 11.4 *Comparison of initial and final glazing options*

Selection of central plant

The extensive thermal analysis of the building led to early establishment of a target cooling load of 12,500kW. Methods for rejecting this heat were then investigated.

There were two principal families of contenders. Those using high level evaporative/air cooled systems and those using sea water drawn from Hong Kong Harbour.

The main disadvantage of the evaporative/air cooled systems was the extensive use of prime floor area at high level (some 2,500m^2) together with the significant use of vertical riser space throughout the height of the building. The disadvantage of the sea water systems was the extensive civil works required to bring the pipework into the site because the Hong Kong Government would not permit closure of the water front roads and the existing sea water pipe serving the 1935 Headquarters could not be satisfactorily refurbished.

The results of detailed cost-in-use analysis showed that the air cooled schemes used some 30% more electrical energy than the sea water schemes (equivalent to about 2,600 US barrels of oil per annum) and this taken with the need to preserve upper level floor space and provide high reliability and flexibility in the heat rejection systems determined that the sea water proposals should be adopted.

Although Hong Kong's heating season is short, winter temperatures can go close to freezing point in the urban areas and a real heating capability is required. Again, the cost in use of various systems was studied and it was apparent that heat reclaimed from the chillers was most attractive. However, it was also realised that the most economic system was highly dependent upon the particular plant selected (size, number and performance of machines).

In order to optimise selection, manufacturers were invited to submit proposals for the generation of chilled water and hot water with seawater as the ultimate heating/cooling medium. Each manufacturer was provided with heating and cooling load profiles for the building (computer generated) and had to support their selection with calculated capital, running and maintenance costs, plus details of space requirements.

One manufacturer submitted three quite different schemes for consideration with present values which differed little more than 5%.

Previous experience in Hong Kong had indicated a higher reliability for separate titanium plate heat exchangers as compared to shell and tube condensers/evaporators. There were also height obstructions which limited the size of machines. These two factors decided the choice.

Choice of air conditioning system

Raised floors fitted throughout general office accommodation are now de rigueur for modern commercial buildings. Under these circumstances the most economic solution logically involves maximising the use of the sub-floor void for distribution and providing almost all services to the occupied space upward from the void.

Floor mounted air conditioning grilles have, of course, have used for many years in computer rooms. However, at the time of concept development for the Bank very few attempts had been made to provide air conditioning in this way for high quality office accommodation. Examination of these existing systems both in the laboratory and at installation sites highlighted two potential areas for concern.

low volumes and discharge velocities, aimed at containing any discernible air movement within the immediate area of the outlet, resulted in little air movement within the room at head height which seemed to give rise to a lack of freshness in the space.

proponents of the systems advocated 100% exhaust at high level which required a space consuming return air system and restricted the architectural form of the ceiling.

In the case of the Hongkong Bank, the Architects were concerned to limit the use of service space above ceiling level. An early option developed to counter the drawbacks of the existing systems used an upstand pillar (or kiosk) which combined the functions of uplighter, high level air conditioning supply and exhaust at low level (Figure 11.5). In parallel with this the Team also developed a personal air conditioning utility which provided air on two levels at each work station and permitted the occupant local control of temperature, velocity and discharge direction. These schemes were taken to a relatively advanced stage with all the principal components being proven. However, it soon became clear that the forest of upstands created by the kiosk system was in direct conflict with the notion of large column-free spaces. Moreover, closer investigation into the previous use of personal air conditioning systems (principally in Europe)

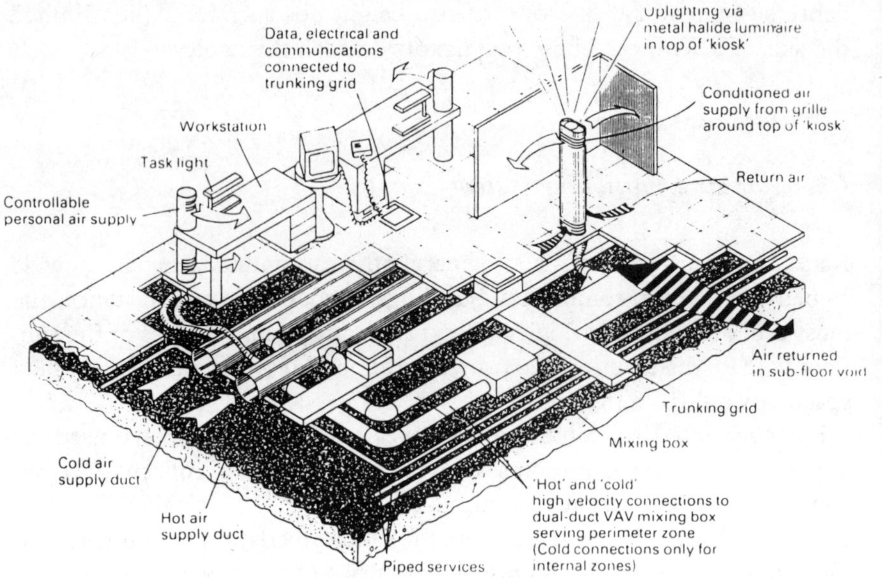

Figure 11.5 *Floor-based services distribution, initial concept (1980)*

showed little, if any, tangible benefit to the occupant apart from the psychological advantage of giving them control of their air conditioning.

From this time forward it was decided to concentrate attention on the development of a general air conditioning system using outlets which were flush with the raised floor. A series of detailed laboratory studies including full sized mock ups (described by Farnfield later in this chapter) led to a final solution using new high duty circular and linear floor outlets combined with a mix of high and low level exhaust.

The early kiosk/personal air conditioning systems had flirted with the concept of dual duct VAV in order to give each work station the option of direct heating. Further detailed economic and spatial studies emphasized the obvious inefficiencies of this approach. The final air conditioning scheme combined a constant volume perimeter supply (scheduled supply temperatures) and a constant temperature VAV system for internal zones (Figure 11.6).

Since the constant volume perimeter supply duct carries cold air in the summer, there is no excess duct capacity for heating and further ducting economies were made by combining the final connections to the perimeter linear outlets. Adequate local control is achieved by conventional air valves in VAV terminal units in the sub-floor — each valve serving up to a

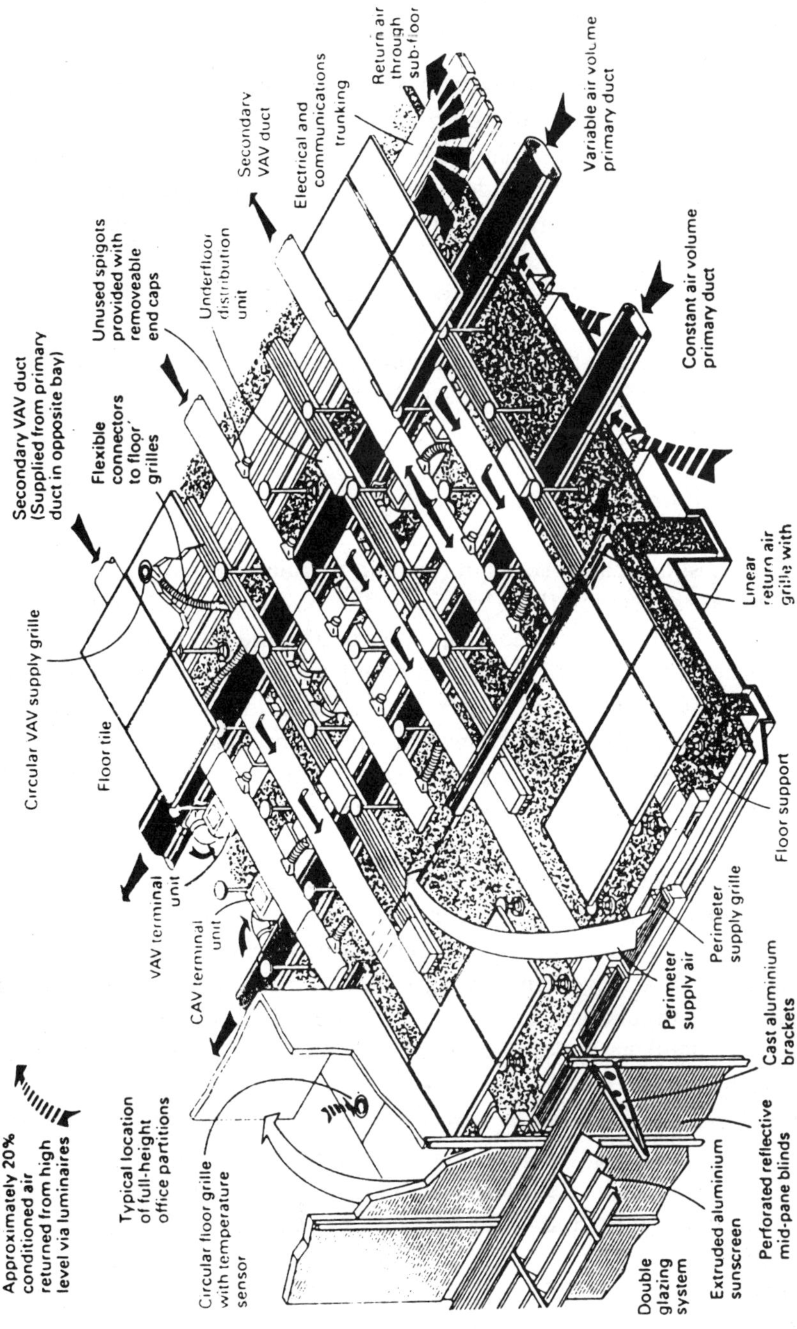

Figure 11.6 Modular sub-floor services layout, as constructed

maximum of five circular floor outlets or two linear perimeter units.

Following discussion with the Hong Kong Fire Services Department it was proposed that the air conditioning system should be capable of providing 100% fresh air to clear a smoke logged floor in the aftermath of a fire. An important feature of this arrangement was that it should be under the manual control of the attending fire officer. The automatic response of the air conditioning system under fire conditions was to shut down — not only to avoid recirculating smoke but also to prevent competition with the staircase pressurization system.

The decision to air condition the lifts and to provide escalator access through the bottom of the Atrium were important considerations in the balancing of air flows for the conditioned spaces. The resulting arrangements are described by Thomas later in this chapter.

Services modules

Figure 11.7 shows the position of the four modular plantrooms and flanking risers on each typical three bay floor. The positioning and arrangement of the units facilitated the modular construction and the proximity of the risers made natural the incorporation of toilet accommodation in the front section of each plantroom. This arrangement provided an excellent buffer zone for noise and vibration, however, the negotiation of pipework and ductwork through a highly serviced sanitary provision and the adjacent structural zone around the masts proved formidable.

The concept of factory packaged building services elements is, of course, not new. However, it has been little used in high rise building apart from in prefabricated bathrooms which are a familiar sight in Japan.

Fundamental to the economics of the modular approach is the degree of standardisation achieved and the approach to replaceability. There is no simple equation linking these factors with the structural and architectural requirements of the building and the final solution is highly dependent upon value judgements.

In the case of the Hongkong Bank it was elected not to make the modules replaceable. Whereas this gave rise to some economies in the structure it resulted in the requirement for ongoing maintenance and replacement to be performed wholly from inside the building. This applied not only to the air conditioning modules on the West side of each floor, but also to the modules containing HWS boiler plant and standby generators at the top of the building.

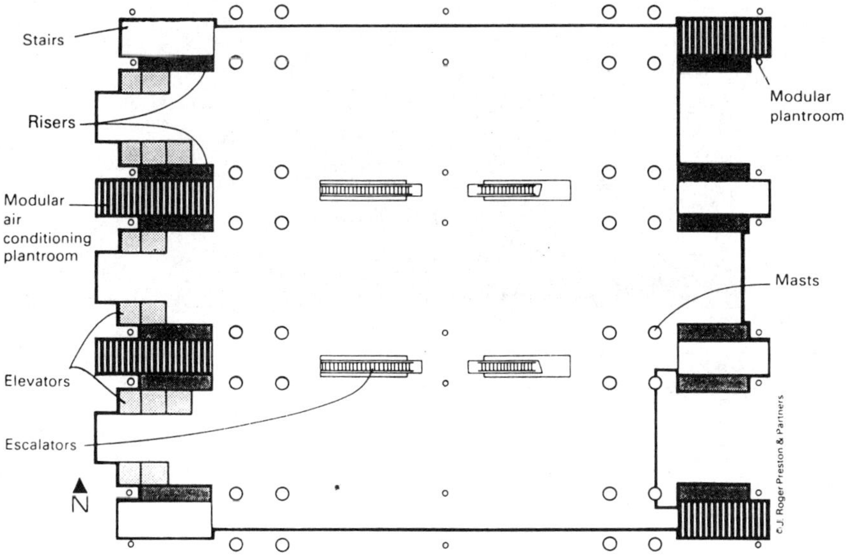

Figure 11.7 *General arrangement of typical three-bay floor*

In general, only the modules on the West-side of the building contain air conditioning plant — the rear of the East-side modules being reserved for domestic water systems, equipment, etc. Groups of air conditioning modules within each seven or eight floor zone of the building are identical. They are individually designed to achieve optimum performance within the space available and there is no doubt that this economy of space could never have been achieved using standard equipment and on-site erection methods.

Prior to commencement of production of the air conditioning modules by the chosen Japanese Sub-Contractor, a full scale prototype air conditioning module was subjected to rigorous testing. This testing included proving the capacity of all components, exercising the control systems and checking for noise and vibration and ductwork leakage. Removal and replacement of all major system components was also checked and demonstrated to the Client's maintenance personnel.

The main features of the final module design are shown in Figure 11.8.

Computerised building management system

Hong Kong has many large buildings with extensive BMS systems and the major control manufacturers are well represented by competent software

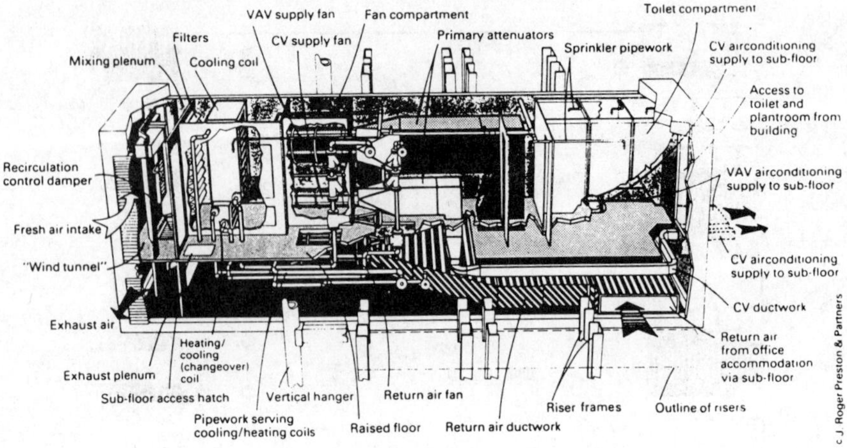

Figure 11.8 *Layout of typical modular toilet and airconditioning plantroom*

and hardware specialists. Thus the Client and Designers were not inhibited in calling for a high performance system for the new building. However, the Designers experience in Hong Kong, as elsewhere, has emphasized the need to match the BMS installation carefully to functional and constructional requirements. Examination of the manufacturers wares and development of the final scheme was thus guided by three important requirements:

1 *Speed of response and reliability:*

Many large systems perform well under light loading but poorly under the very heavy loading generated by any real emergency.

2 *User friendliness and flexibility:*

These aspects are largely controlled by software characteristics which are specific to individual manufacturers and difficult to modify.

To improve these elements the Team investigated numerous features to simplify operator/machine communication (these included audio control, dynamic graphics, large scale video projection, etc.) although few of these features were found to be suitable for final incorporation. More fundamental requirements were those of full field programmability for control algorithms and the involvement of senior client personnel in the development of referencing systems and reports to ensure that the computer related to the Client's established methods of operation.

3 *Separation of local control from BMS control:*

In recognition of the fallibility of any computer systems and the need to locally control and commission systems off-line, the Designers insisted on all logical groupings of plant having local hardwired control with a clearly defined interface with the computer system. This was particularly imported in view of the large number of different sub-contractors working under the fast-track regime.

The process of development of the BMS and the major features of the final installation are described at the end of this chapter, by Yip.

Commissioning

A high degree of self balancing was automatically achieved in the office air conditioning primary ductwork by the use of CAV and VAV terminal units. To meet the requirement for flexible positioning of the air conditioning outlets, the secondary distribution system was also made generally self balancing by assigning the floor outlets and flexible duct connections a relatively high resistance compared with the flat oval secondary distribution header section.

Commissioning of the building as a whole was carried out under a separate sub-contract. However, it was apparent that whereas this could deal with the mechanics of systems balancing etc., it was not of itself sufficient to ensure acceptability of system performance — for which detailed design knowledge was clearly required.

In advance of commissioning and documentation, the Consulting Engineer prepared extensive computer equipment listings and systems performance descriptions — totalling some thirty volumes in all. These were handed to the Sub-Contractor who then worked through the commissioning procedures alongside the Consultants own experienced engineers who ensured close adherence to the design intent.

Inevitably, commissioning commenced somewhat later than had been programmed, but the detailed pre-planning provided a sound basis for the work and no major difficulties have been experienced.

Full systems performance tests will take place during the Summer of 1986 under the supervision of the Designers.

User response

The user took full occupation of the building, as programmed, between July and November 1985. It is thus too early for a full assessment of the Building services systems performance. Nevertheless, to date the Client has expressed himself well satisfied with his new building.

It has been especially gratifying to the Designers to note the performance of the floor based air conditioning system. Temperature control and air distribution have proved entirely satisfactory even in departments where there is a high density of pre-dominantly female occupants.

Evaluation of the building envelope: *Martin S. Ratcliffe*

During the concept development stage, combinations of external shading and glazing had been examined. An efficient external shading had been agreed on and designed to allow greater freedom in the choice of glazing, the most promising of which appeared to be reflective glass.

The fast-track nature of the project demanded that the air conditioning systems to office floors be sized at an early date. Cooling load calculations were accordingly based on the use of the reflective glass. Unfortunately, this glazing did not meet with the Architect's approval on closer scrutiny. An alternative, therefore, had to be found.

This section outlines the processes involved in the reselection, with particular reference to the method of communicating the thermal requirements to the Architect.

Key requirements

The building is virtually 100% glazed on the major facades which look North and South, providing possibly the most sought after views in Hong Kong. The profile of the ceiling is designed to maximize the view out by increasing the internal floor/ceiling height at the perimeter of the building to 3.45m, which is fully glazed. Carefully designed external shading helps reduce the impact of solar radiation, but this extent of glazing combined with the high outdoor air temperatures (33° C) necessitates good thermal performance if reasonable cooling loads and comfort temperatures are to be maintained.

To emphasize other major features of the building, the Architect was very keen to utilise clear glazing externally. Solar control had therefore to be provided in the main by an inner skin, with an inherent loss in relative performance. This soon became known as the back-up wall.

The limits on cooling loads which could be accommodated had been set by the early selection of air conditioning plant, and provided the basis for a set of criteria for thermal performance. This could then be compared and adjusted as necessary with that of comfort. Whilst thermal comfort is the result of heat transfer processes in a similar manner to the load imposed on an air conditioning system, its dependence on position requires that it is calculated separately. (The CIBSE Admittance Method for the calculation of temperature and load does yield a dry resultant temperature, but this is an average value only and is not necessarily applicable to adjacent windows.)

Comfort temperature

Comfort (or more properly dry resultant) temperature (1) at the perimeter of a glazed building is influenced by air velocity and temperature, surface temperature of walls, glass etc., and by the intensity of any solar radiation which passes through the glazing and impinges on the occupant. A design air temperature of 24 deg C had been chosen and a comfort temperature of 27° C was agreed upon (2) as being the maximum which could reasonably be tolerated and which was considered achievable in practice.

The dependence upon radiant energy means that dry resultant temperature varies with position. The closer that position to a warm surface (e.g. window heated by solar radiation) the greater the influence of that particular surface. The radiant component of dry resultant temperature must therefore be calculated using the laws of radiative heat transfer for each position of interest within the room. This is normally done by using shape factors.

Solar radiation which passes straight through the glazing and impinges upon the occupant is also a radiative heat transfer process, but it is far easier to determine the effect on dry resultant temperature than that of surrounding surfaces. Being in the form of a parallel beam, the effect is not diluted with distance so that provided the beam actually hits the occupant, distance from the glazing is irrelevant. (A simple rule of thumb is that every 25W/m^2 of transmitted solar radiation results in a rise in mean radiant temperature of 1° C. About 0.5° C dry resultant temperature.) P.G.T. Owens (3) provides further details of these calculations.

Just as in a domestic heating system where radiators are located below windows to offset 'cold radiation draughts', so the effect of warm glass surfaces can be reduced by the positioning of the air conditioning supply grilles. This approach was employed on the Hongkong Bank project. Perimeter floor grilles direct air across the face of the glazing, providing warm air in winter and cool air in summer. Invariably this leads to increased heat transfer rates and hence energy consumption but does increase the useful floor area.

Solar gain factor

The CIBSE Admittance Method uses a set of factors to calculate the solar cooling loads of glazing, and the CIBSE Guide provides such factors for a number of typical glazing systems. The method by which these factors have been determined is explained in Jones (4).

As the possible solutions for the back-up wall went beyond the typical systems for which the CIBSE had pre-calculated solar gain factors, it was decided to write our own computer program based on Jones (4) above. This allowed us to examine virtually any combination of glass and blinds.

Transmittance/absorptance charts

The ingress of solar radiation is governed by the transmittance, absorptance and reflectance of each component, the geometry of any blinds and heat transfer coefficients. The latter are affected by the combination used, but the others are fundamental properties (and easily obtained from the manufacturer.) Using such properties it was possible to present the Architect with a set of criteria which accurately specified the thermal performance whilst being in a meaningful form.

This method proved highly successful. Rather than laboriously examining individual options, a few general arrangements could be indicated on a transmittance v absorptance chart. The architect could then simply select components which fell within the acceptable range. An example chart is shown in Figure 11.9.

The chart is based upon a double glazed solution where the outer pane of glass is specified. (In this instance 10mm clear glass.) The line represents the limiting values for combinations of solar transmittance and absorptance. (For any value of transmittance, t, and absorptance, a, reflectance is given by the expression 1–t–a. The chart can therefore be drawn with any

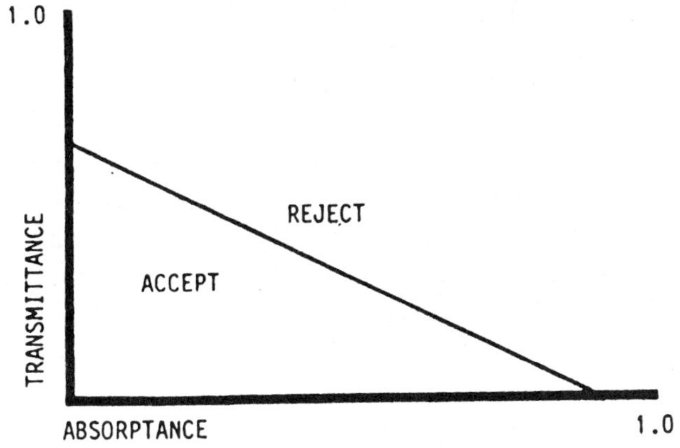

Figure 11.9 *Transmittance/absorptance chart*

two of the three properties.) The chart was constructed by determining solar gain factors for the double glazing using the fixed properties of the outer glass and various combinations of transmittance and absorptance for the inner glass. The values of solar gain factor were then used to calculate cooling load using a computerised version of the CIBSE Admittance Method. By trial and error, values of transmittance and absorptance were found that just met the cooling load criterion.

Provided these did not result in too high comfort temperatures, they were used to plot the curve as shown. In practice, this meant that high overall transmittance was not allowed. High absorptance, leading to otherwise high glass surface temperatures, was offset by the use of the perimeter supply air grilles.

Similar charts were constructed for several other arrangements, e.g. double glazing with specified blind and outer glass. In general, one component of a glazing system can be represented in this way provided all other components are specified.

Final selection

Eventually, the choice was narrowed down to three systems of glazing. All consisted of double glazing with clear outer glass.

The first used a novel inner pane containing louvres actually within the glass thickness itself. During manufacture the glass was dosed with a photosensitive chemical. This turns opaque in light; the degree of opacity depending upon the exposure, rather like film negative. Virtually any pattern could be produced by suitable masking. A fixing agent was then added.

This glass was said to owe its origin to the US space programme. It had never been used in commercial buildings before. Extensive tests were carried out by an American research organisation to ascertain the solar properties for various patterns and densities. The transmittance v. admittance chart proved invaluable.

The second option used a more familiar reflective glass for the inner pane.

The third employed perforated mid-pane blinds and a lightly heat absorbing inner pane.

There was some reluctance to accept blinds as the major solar control element. It was feared that these would not be used as intended and lead to complaints of discomfort. An agreement was reached whereby the occupant generally could not raise the blind, and could only adjust the slat angle between certain limits. The contribution made to thermal performance by the heat absorbing glass was minimal and was introduced to reduce glare when the blinds were partly opened.

User's choice

Costs were estimated for each scheme and full size mock-ups made. The client was then invited to make the final decision. He chose the third option.

Initially, it had been intended to employ the same glazing system on all facades of the building, but adjusting the performance downwards if possible to suit each facade. Thermal performance criteria had been based upon the Southern facade, being generally the more onerous. It was found that for this particular arrangement it was just possible to omit the blinds to the North facade and use photoelectric switching of artificial lighting; taking advantage of the higher daylight factors and in the process reducing internal gains.

Integrated vertical transportation system: *K. Thomas*

The design criteria for the lift systems in the building was based on the requirement to provide lift service to 17% of the population per 5 minutes during peak traffic periods with an assumed population of 6100 people. The target waiting interval during peak periods was set at 25–30 seconds.

Vertical transportation proposals

During the initial design of the Building numerous alternative vertical transportation schemes and patterns were studied. These were all variations of the following principal schemes:

- single deck lift scheme
- combined scheme with single deck + double deck lifts with sky lobbies
- combined scheme with single deck lifts + escalators.

The single deck lift schemes resulted in a requirement for approximately forty lifts to meet the design criteria. This was rejected by the Design Team on the basis of the high percentage of plan area used not only for the shafts but also the associated lobby areas.

The combined schemes using double deck lifts theoretically provided a close solution to the design criteria. However the reduction of usable car floor area caused by increased car frame and safety gear sizes made this scheme uneconomic.

After extensive studies of traffic flow, including discussions with an independent office management consultant, a combined scheme of lifts and escalators was proceeded with. The final concept uses high speed (6 m/s) lifts serving double height reception areas. Twenty four passenger lifts in a configuration of six banks of four were used. The use of double height levels as sky lobbies from which passengers transferred to escalators for transportation to adjacent single height floors co-ordinated well with the Architects planning proposals. The escalators were located in the core of the office accommodation thus enabling the spaces between lift banks at single height floors to be utilised to suit departmental needs. Goods lifts were located in each of the four escape stair cores. Two of these lifts doubled as designated firemen lifts to satisfy Hong Kong Fire Services Department requirements.

For the Banking Hall it was decided that access from the street level plaza would be best served by escalators to achieve the substantial traffic flow. The escalators pass through the glazed base (underbelly) to the air conditioned atrium. This posed a number of air conditioning problems but provided dramatic public access to the Banking Hall.

Lift car design

The Architect wished to express the mechanical elements of the lifts which would be seen internally through glazed shafts in the atrium or externally through complete glazed elevations. During the design period the Sub-Contractors Works were visited to agree modifications to doors, steelwork and the like, to visually clean up the car appearance. Items of equipment allied to safety were not changed. The cars were constructed from a Jansen hollow steel section from Sweden and a full analysis of the construction was undertaken by the University of London to ensure that safety gear application loads were achieved. The car walls were manufactured in Japan from a toughened glass/fibre blanket/paraline glass sandwich. With internal illumination this achieved the effect of a Japanese Shoji paper screen. A prototype car was completed prior to manufacturing approval being granted by the Design Team. The cars are air conditioned using roof mounted low profile packaged units.

A micro-computer control system was installed to perform all functions of safe lift motion and door control. Each lift car has its own controller incorporating a 16 bit microprocessor mounted on the car with main group controllers located in an associated machine room. Continuous monitoring of lift groups is carried out to ensure prompt passenger service.

Traditional motor generator sets were replaced with silicon controlled rectifiers so that power is only consumed when the lift is moving. This has resulted in an estimated 30% saving in energy used.

Escalator design

Design development for the escalators ran in parallel with that for the lifts. As with the lifts the Architect wished to express the mechanical elements of the escalators. This resulted in the truss sides being glazed and the appearance of steelwork, brackets, etc., being cleaned up.

All escalators above the Banking Hall are controlled by pressure pad operation to reduce energy usage and consequential loads on air conditioning installations. Direct-on-line starting was required to achieve satisfactory operation and special permission had to be obtained from Hongkong Electric to move away from star-delta starting.

The escalators serving the Banking Hall from the plaza posed a number of design problems. With the plaza being a dedicated public access the Architect did not want intermediate support for these escalators. The eventual truss design consisted of square tubular sections of adequate strength to withstand its full contract load (9,000 people per hour) and three second gust typhoon wind conditions. The escalators had to be fully weatherproofed including the glazed truss sides.

Impact on air conditioning design

The design of the buildings air conditioning systems had to incorporate certain features as a result of the form of vertical transportation used. These included the impact on the atrium air conditioning systems of large holes in the roof and underbelly through which the escalators passed, and the need to maintain acceptable conditions in the glass sided lift-shafts.

Atrium pressurization

The atrium is some 39m high rising from the level 3 Banking Hall to the slab of level 13. It is open to surrounding office accommodation and consequently has a volume of 120,000m^3. The escalators entering the Banking Hall from the plaza through holes in the glazed atrium underbelly caused concerns related to maintenance of satisfactory environmental conditions. The potential problems were compounded by escalators passing through the level 13 slab at the top of the atrium and by the large number of lift door openings into shafts up to 44 floors high.

The University of Western Ontario had been commissioned to carry out aerodynamic tests in their wind tunnel on aspects of the structure. These were extended to examine the effects of the holes in the underbelly with pressure differentials generated by the building's wind microclimate. Using a mix of scale and mathematical models it was predicted that pressure variations across the underbelly would cause 'slugs' of air to alternatively enter and leave the atrium through the escalator holes at a frequency related to prevailing wind conditions. It was concluded that, to prevent undesirable effects due to the escalator holes, pressurization of the atrium was required. Under normal maximum reference prevailing wind conditions an excess air outflow through the holes of 10m^3/sec was required. To achieve this, 100% fresh air supply plants with a total capability of 8m^3/sec were added. Under lesser ambient conditions lower excess air flow rates are required. The escalator holes at the top of the atrium (where traffic flows were not high) are effectively closed using glazed barriers with revolving doors.

Pressurization is achieved in a number of stages. The module air conditioning systems introduce fresh air to the office accommodation at a constant rate. The balance between this fresh air supply and that extracted via toilet systems, etc., contributes as a base stage. To reduce running costs, the atrium pressurization plants are brought into operation as further stages when adverse pressure differentials are sensed across the underbelly. This process is monitored by the Building Management System and can be adjusted to suit operating experience.

No information on the performance of the systems is available as yet.

Lift shaft ventilation

Local regulations required that the lift shafts be naturally ventilated with high level smoke exhaust and low level air intake. The adoption of single glazing to the external, west facing walls of the outer shafts resulted in appreciable solar gains. Calculations suggested that this would result in unacceptably high temperatures within the shafts with consequential gains to the glazed cars and operating losses for the lift car packaged air conditioning units. Mechanical extract systems are installed to operate as an alternative to natural ventilation when temperature conditions at the top of the shaft dictate.

On floor services for new office technology: *Trevor C. Farnfield*

With a projected services life of 25 years it was considered essential that the on-floor user end of the installations was flexible for future planning and adaptable to future technology.

The choice of air conditioning system evolved with a number of key objectives in mind. These were:

- low energy usage
- good local control
- ease of maintenance
- flexibility

Additionally, construction factors were considered. Prime amongst these were the need to achieve consistent high quality and speed of erection.

Selected air conditioning system

A number of air conditioning systems were considered with the final selection being a single duct variable volume system serving the internal zones and a mixed constant/variable volume system serving the perimeters. The air is discharged into the occupied space from the raised floor and returned via outlets at low and high level. The subfloor distribution systems were prefabricated off site in a modular form.

The structural solution for the building resulted in a subfloor void with a nominal clear height of 600mm. This reduces to 225mm adjacent to the primary beam zones. From the two centralised plant modules on the west side primary air distribution is East-West with secondary distribution North-South within the structural bays. These principles are paralleled for the electrical services contained in the sub-floor void.

The available sub-floor void was limited and detailed planning and co-ordination was required of the Design Team to maximise its use. A 100mm nominal zone was reserved immediately below the 1200mm square floor panels for the air conditioning and electrical outlets together with their associated flexible connectors. A regular services pattern evolved which not only incorporated all required elements but allowed for simple maintenance access. This pattern was developed further by the Sub-Contractor into prefabricated linking assemblies with a common sub-frame support system. With floor areas up to 2,700m^2, it is worth noting that the Sub-Contractor maintained a programme which required him to complete all services on a floor every 22 days.

Development of supply outlets

Having settled on a floor based air conditioning system, the Design Team looked at similar VAV systems in Europe. The use of commercially available outlets had to be rejected for this project for the following reasons:

- the subfloor void was to be used for return air and local fire regulations would not permit the use of plastic materials under such circumstances.
- much higher occupancy rates were anticipated in Hong Kong; the European installation had typically 12–15m^2 per person whilst 6–9m^2 per person is common in Hong Kong.

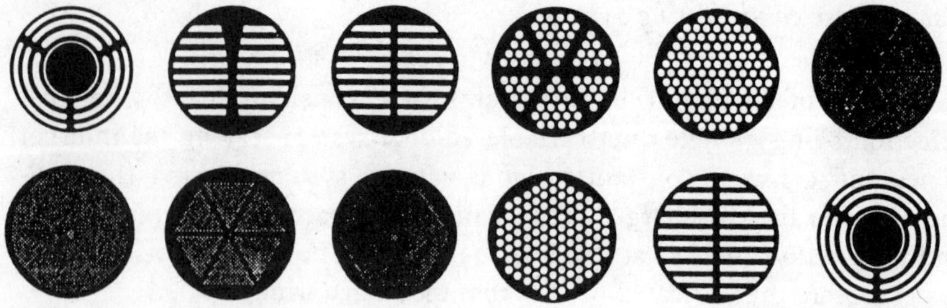

Figure 11.10 *Alternative patterns considered for face of floor mounted airconditioning diffuser*

- higher supply/space air temperature differentials were required to achieve the objective of efficient energy usage and compactness of central plant.

It was concluded that a high duty circular outlet needed to be developed for central zones but that a linear format was more appropriate at the perimeters. On this basis a full size test rig room was set up by a manufacturer in Germany and laboratory development proceeded.

A large number of circular diffuser patterns and sizes were examined. (Figure 11.10) Key factors for satisfactory performance were a large effective grille area and the achievement of rapid room air induction. Optimum results were achieved using a 200mm dia. unit with the grille face formed from concentric rings around a solid core. A set of internal radial vanes were used to induce swirl in the emerging jet. The final form of the outlet used on site achieved satisfactory air distribution with volume flow rates of up to 35 litres/sec. For planning purpose outlets were rated at a nominal duty of 30 litres/sec thus allowing some margin for transient loads.

For the perimeter linear outlet a plenum/grille combination had to be developed which combined the variable and constant volume systems. Key factors for defining satisfactory performance were the maintenance of discharge velocities throughout the VAV range, the generation of secondary circulation by induction effects and the control of dumping. These factors are inter-related. In the adopted solution, a balanced damper bypass system is used to differentiate constant and variable supply components; the former being discharged adjacent to the glass face. Rapid entrainment of VAV and room air is satisfactorily achieved.

Return air

The series of laboratory tests were extended to investigate differing locations and percentages of room exhaust air. The following parameters were considered:

- temperature gradients.
- return air temperatures.
- subjective feelings of comfort.
- effects of smoking.

It was concluded that comfort conditions were satisfied if a minimum of 15% of the return air was drawn from above the occupied zone; the remainder returning elsewhere. The results of these tests were applied to the use of a plenum return air path comprising ceiling luminaires, floor grilles, ceiling voids, internal cladding and floor voids. The pressure distribution through the proposed return air path was established from model tests undertaken by BSRIA.

Controls

With flexibility having been achieved for internal zone outlet locations, consideration was given to locating control sensors such that they could also be moved easily to suit planning needs without affecting architectural components.

It was found from laboratory tests that by utilising the small pressure differential between the occupied space and the subfloor void, air could be drawn across a sensor located in a floor mounted outlet. Each type of outlet was developed to incorporate a port for this facility. As installed on site the majority of sensors are located within linear return air grilles on the primary beam lines.

For double height areas of the building where the floor void was not used for return air an outlet incorporating a fan assisted sensor was developed.

Electrical services

In parallel with the development of the air conditioning outlets work was also progressing related to electrical services distribution. With the office of the future in mind the prime objectives were to enable power, communications and data systems to be fed to work stations in such a fashion that new or relocated workstations could be plugged into grid wired distribution networks with minimal disruption and effort. As with the air conditioning, primary distribution was East-West with secondary distribution North-South. Flexibility was achieved using two key elements which were the underfloor distribution units (which link into the secondary trunking distribution) and floor mounted outlets. The latter are identical in appearance to the air conditioning type. They are flush mounted within the raised floor panel and contain five segregated compartments. Each compartment can handle a number of different functions by use of different segment plates. These outlets connect to the underfloor distribution units by flexible conduits and are linked above floor into a furniture cable management system. Special variations of electrical outlet were produced to feed terminal computers and to satisfy the requirements of difficult data intensive departments such as the Foreign Exchange.

Departmental planning

With the aid of a computer aided drafting system the Design Team produced complete planning layouts for five thousand workstations on thirty four levels of the building. A Prime 750 computer was used in conjunction with a software package called DOGS (Design Office Graphical System). This package handled all aspects of the planning process including furniture and workstation layouts, partitioning layouts, subfloor services connections and soffit services. It is now the Bank's intention to use this database, which was updated to as built status, for future departmental replanning exercises.

The future

The Design Team confidently believe that the Client has in his new headquarters the facilities to react quickly and efficiently to the changing needs of new office technology. Inevitably with flexibility came complexity and additional capital costs. However, there is no doubt that the Client will reap the benefits of forward thinking and his capital investment in the years to come.

A new concept in high speed building management systems: *Patrick Yip*

The Building Management System (BMS) for the New Headquarters is unique, not because of its size and complexity, but in its design philosophy and implementation. The successful application of a BMS is a direct result of a methodical structured approach, an appreciation of the technical and commercial procedures involved and a willingness to adopt new technology, ideas and operational methods.

Design brief

Direct interaction with the Client's Team has been an important element in the development of the Brief for the BMS. The Client specifically laid down his needs through consultation with his user departments and a firm statement of requirements was communicated to the Design Team.

In seeking the best in control and management for the building, the Client recognized the need for an up-to date state of the art assessment. The particular problems of such complex installations in the Hong Kong environment were recognized with the result that the Client was in agreement with the Design Team that the final system be developed in close co-operation with a selected manufacturer.

Tendering stage

There were some fifteen companies worldwide claiming to offer a computer based building control system. Nine potential suppliers of computerized BMS were assessed. Assessment was carried out through visits to the suppliers principal centres of operation, witnessing system demonstrations and discussion with key commercial and technical personnel.

It was concluded that the current generation of BMS systems have many common features in their general modular approach to both hardware and software. However, there are important differences in detailed capability and approaches in the key areas of data gathering panel (substation) performance, user programmability and Hong Kong experience/ service.

This thorough analysis of commercial, practical and technical factors led to the prequalifying of four firms. Whilst the Design Team was satisfied that all four firms had the technical and commercial capability to complete a satisfactory installation, none of them offered as standard all the desired features for this project.

The BMS package was tendered as a performance specification with a provisional bill of quantities. This allowed the Design Team to progress the package before details of all building services components were known. The general concept was sufficiently advanced to produce a specification of adequate integrity. Due to the complexity of BMS interface encompassing all services installations in a project of this size, great care was taken in producing a comprehensive and compatible set of documents. The following basic parameters were fully described in the Tender Documents:

- Size and capacity of central system
- Spare capacity required for components
- Performance of system in event of component breakdown
- System response time
- Definition and demarcation of field interfaces
- BMS interface with M & E systems
- Special feature requirements
- Equipment standards and codes of installation
- Testing and commissioning requirements (on site and off site)
- On-going maintenance agreement.

The adoption of this approach provided the tenderers with the opportunity to express their comprehension of the particular project needs and their expertise in formulating design solutions. This played a major role in the subsequent tender analysis by the Design Team and the ensuing recommendations made to the Client.

Interim stage

A sub-contractor was appointed to enter into an Interim Agreement. This involved a design development period during which time the requirements of the BMS were more precisely defined with the objective of configuring a more efficient and cost effective design. Careful scrutiny of all the building services installations and their relationship had resulted in the emergence of a clearer philosophy in the utilisation of the BMS. This had come about due to close co-operation between the Design Team, and the Sub-Contractor and the Client in the formalisation process. Each point of interface was analysed in an attempt to make full use of the facilities available in an advanced computerised system.

Final system

The installed system is based on distributed processing from hot redundant central computers. Each of the two mini computers includes a 10MB dual port fixed head disk. The on line mini updates the standby as changes in the field occur. The standby automatically takes over in an updated condition should the on line computer fail.

The approximately 6500 control and monitoring points in the building are connected to the central computers through two levels of distributed processors.

The lowest level is the Field Processor Unit (FPU) which connects directly to field inputs and outputs. This device polls up to 16 points. It detects all changes of binary inputs, digitizes all analogue inputs and converts digital commands to analogue or binary outputs.

The next level of processor is called a Field Interface Device (FID). The FID polls up to 16 FPU's (about 250 points). The FID picks up all changes in polled FPU status or analogue points and passes them on to the Central Processing Unit (CPU). It also distributes any commands from the central processor to the proper FPU. In addition the FID has considerable programming capability within it. It is down loaded with programmes from the CPU and should the central system fail the FID continues to carry out all programmes down loaded into it.

The CPU communicates with the FID via standard telephone grade COM channels with full duplex modems. One COM channel can handle up to 16 FID, but only 4 to 6 are being used to ensure fast response. An alarm sensed at the FPU will be displayed and printed out at the CPU I/O within 2 seconds under normal conditions.

The BMS polls each of its 6500 points within 0.5 second and distribute a command to a point at the same rate. While computer internal processing speeds approach this capability with internal batch processing, the time delays involved in communicating with field interfaces make central only processing unacceptably slow on large systems.

The distributed processing acts as a filtering process that relieves the central processor of routine operations and leaves it only those tasks which it does best.

The central processor receives only changes. It outputs only operator commands or program commands that cannot be handled by the distributed processors. Its large memory allows it to do calculations based on two or more inputs. It has the capacity to generate logs, summaries, and dynamic colour graphics from the data on its disks.

The central computers and the main control room are located on the 27th floor. A repeater control room is located on the second basement level mezzanine overlooking the building main plant.

Reliability was one of the key requirements for the system. At the time of selection direct digital control was discounted as being not sufficiently advanced. The decision was made to allow all plant to be run independently of the BMS with the BMS being capable of providing limited control functions.

Energy management potential

The Building Management System software and hardware has been tailored to carry out an extensive range of functions to reduce energy consumption, generally through its ability to gather information at a central point to assist the maintenance staff, and to highlight faults and inefficiencies that would not normally be discovered in good time.

The main features of the energy reducing potential of the BMS are summarised below:

1 Each air handling module can be independently controlled to suit occupancy times, using optimum start/stop software. This allows any floor to be used without running plant to unoccupied floors. The BMS measures outside air and inside air temperature on each floor. The optimum start-up/stop time is calculated using specially written software, for both heating and cooling modes.

Proprietary optimum start packages using micro-processors have a limited ability to learn by experience. By comparison, the BMS system installed is able to monitor and log events and conditions which, together with the facility to easily change constants (e.g. time constants) enables the software to fine-tune the performance of the building and its installations.

2 In the event of a particular floor out being occupied, the BMS can be used to run the air conditioning if required to maintain reasonable temperatures, but without any fresh air supply.

3 Each air handling module provides constant volume variable temperature supply air to the perimeter to offset conduction gains/losses through the glazing. This temperature is scheduled to outdoor air temperature by a simple linear relationship. This schedule can be changed remotely by the BMS for each individual module.

4 Refrigeration plant is brought on line, ready to run, by the BMS by comparing the chilled water requirement to the chilled water output at frequent intervals. The COP of each machine is also monitored.

5 Lighting and small power consumption is monitored on each floor to provide a basis for departmental or tenancy charging. This should act as an incentive to reduce energy consumption.

6 The BMS monitors and totalises all incoming power. Present and predicted power consumption is displayed. Standard software can predict a specified maximum demand being exceeded. In such an event, non-essential plant and equipment can be shut down. These are restored automatically on sufficient fall in demand.

Preventative maintenance

A separate maintenance mini computer which is compatible with the Building Management System computers continuously receives data from the management system pertinent to maintenance such as total running time of equipment. The maintenance system is programmed to print out preventative maintenance work orders on all equipment either on a calendar or running time basis. Work orders can include such data as parts and tools needed, estimated man hours and costs.

If a work order is not acknowledged as completed in a specified time, it is repeated. The maintenance system is also capable of maintaining inventory of the thousands of spare parts for the building, and will issue restock reminders when a stocked item falls below minimum quantities.

The maintenance computer may also be used to develop new programmes for the BMS. This feature allows off line work on the management system software without taking one of the redundant computers off line for lengthy periods. Tape decks are provided for loading and dumping software, but the high speed random access required for BMS operation precludes their usage as on line devices.

Works testing

One of the major means of ensuring the performance of the BMS was the specified stringent requirements for Works testing.

Given that there were special features which the Supplier had to develop for the project, factory demonstration was considered to be essential. The time constraint for installation and commissioning of the central and front-end equipment made it vital to carry out as much pre-checking on the system as possible. Evaluation of system response time and the performance of the central system are also best carried out in factory conditions.

Testing of the system was carried out prior to delivery of the central system. The total central system was available fully set up with representative FID lines and FPU's. The system was connected for field simulation and tests were carried out for the following:

- Response time
- Redundancy
- kWh totalization
- Sea water interface
- Chiller interface check
- Air Conditioning module simulation tests
- Constant volume supply air reset
- Graphics
- Programmable techniques

Not all the tests were initially successful which confirmed the benefit of factory testing in identifying problems before they appeared on site.

Off site installation

In line with the general concepts for the Building, much work for the BMS was carried out off site. This included the complete BMS wiring for all plants and sensors in the modules. Thorough operation checks also took place in the Works before the modules were delivered to Hong Kong to ensure all interface connections were correct, thus reducing commissioning time on site. A similar exercise was carried out for all switchboard and motor control panels in which control interlocks were simulated and checked for accuracy before field wiring commenced.

Point schedule

The use of a point schedule has proved particularly useful for co-ordination with other systems and sub-contracts. Not only was it used as a reference for all system controls and their relationship with BMS but was also heavily used for commissioning. All points were checked for correct field wiring and calibration verified before being connected to FPU's and the main system. In parallel the central system was rechecked after setting up on site. As the system began to grow, a full team of the Sub-Contractors's engineering staff carefully monitored the performance of the system. The size and complexity of the project inevitably meant there were a number of changes on site. Interface details were modified to suit plant requirements. New points came about as a result of additional requirements from the Client.

Installation success

The success of the BMS installation required flexibility in the design and construction approaches. The ability of the Sub-Contractor to respond to changes and the ability of the BMS system to be programmed on site also played important roles. Equally, an understanding Client, who has always been willing to participate in the decision making process in a positive way, added to the successful story of the Building Management System for the Project.

References

1 CIBSE Guide. *Criteria for Design*. Institution of Building Services Engineers (UK) Section A1, 1978.
2 Ratcliffe M. and Tuddenham D. *A study in relating air-conditioning and glazing performance: the new Hongkong Bank headquarters*.
3 Owens P.G.T. *Air conditioned comfort and sunshine*. JIHVE, Vol. 37, July 1969.
4 Jones R.H.L. *Solar radiation through windows in theory and equations*. Building Services Engineering Research and Technology Vol. 1, No. 2, 1980.

Index

In the following index, 'air conditioning' is abbreviated to 'a.c.'